From agricultural to food law

Also available in the 'European Institute for Food Law series':

Fed up with the right to food?
The Netherlands' policies and practices regarding the human right to adequate food
edited by: Otto Hospes and Bernd van der Meulen
ISBN 978-90-8686-107-1
www.WageningenAcademic.com/righttofood

Reconciling food law to competitiveness
Report on the regulatory environment of the European food and dairy sector
Bernd van der Meulen
ISBN 978-90-8686-098-2
www.WageningenAcademic.com/reconciling

Governing food security
Law, politics and the right to food
edited by: Otto Hospes, Irene Hadiprayitno
ISBN: 978-90-8686-157-6; e-book ISBN: 978-90-8686-713-4
www.WageningenAcademic.com/EIFL-05

Private food law
Governing food chains through contract law, self-regulation, private standards, audits and certification schemes
edited by: Bernd van der Meulen
ISBN: 978-90-8686-176-7; e-book ISBN: 978-90-8686-730-1
www.WageningenAcademic.com/EIFL-06

Regulating food law
Risk analysis and the precautionary principle as general principles of EU food law
Anna Szajkowska
ISBN: 978-90-8686-194-1; e-book ISBN: 978-90-8686-750-9
www.WageningenAcademic.com/EIFL-07

The enforceability of the human right to adequate food
A comparative study
Bart F.W. Wernaart
ISBN: 978-90-8686-239-9; e-book ISBN: 978-90-8686-791-2
www.WageningenAcademic.com/EIFL-08

EU Food Law Handbook
edited by: Bernd van der Meulen
ISBN 978-90-8686-246-7
www.WageningenAcademic.com/foodlaw

efla
european food law association
association européenne de droit alimentaire

From agricultural to food law

The new scenario

edited by:

Vicente Rodríguez Fuentes

Proceedings of the European Food Law Association 19th International Scientific Congress

Wageningen Academic Publishers

EAN: 9789086862511
e-EAN: 9789086868018
ISBN: 978-90-8686-251-1
e-ISBN: 978-90-8686-801-8
DOI: 10.3920/978-90-8686-801-8

ISSN 1871-3483

First published, 2014

Wageningen Academic Publishers,
P.O. Box 220,
NL-6700 AE Wageningen,
The Netherlands.
www.WageningenAcademic.com
copyright@WageningenAcademic.com

Table of contents

About the authors

Ferdinando Albisinni

Fernando Albisinni is full professor of European Food and Agricultural Law, Università della Tuscia, Viterbo; Professor Jean Monnet of European Food and Agricultural Law; and Professor for the College of Europe of Bruges in programs on Food and Agricultural Law with Universities and Public Institutions in Poland, Serbia, Montenegro, Macedonia. Member of the Italian delegation to the O.I.V.-Organisation International de Vine, Paris; Expert of the Italian Federation of Consortia of quality wine producers. Member of the scientific committee of the Italian Ministry for Agriculture, Food and Forestry, for the simplification of agricultural and food legislation; National Secretary of the Italian Food Law Association; Council Member of the European Food Law Association; Director of the legal Data Bank on Food Law *'Diritto alimentare. Mercato e sicurezza'* (www.leggiditaliaprofessionale.it); Vice-Director of the legal journal on line *'Rivista di Diritto alimentare'*; Member of the scientific board of the legal journal *'Diritto e giurisprudenza agraria, alimentare e dell'ambiente'*; Ordinary Member of the Academy of Georgofili, Florence; Member of the Italian Association of Comparative Law; Member of the Institute of Comparative and International Agricultural Law; Member of the Bar of Rome; admitted before the Supreme Court.

Alessandro Artom

Alessandro Artom is attorney at law before the Court of Appeals Milan and the Superior Court Rome; He is founder and partner of *'Studio legale associato Artom – Papini'*, law firm dealing with the regulation of food and beverages under EU and national law and the EU and national law regulating production and sale of consumer goods. In this regard, the law firm assists the major national and international food/beverages companies operating in Italy. He is a member of Agrofood Legal®: Agrofood Legal network constitutes a free, close and personal cooperation by the participants through respective law firms located in Rome, Milan, Frankfurt and Seville, in order to supply with a qualified and specialized legal professional assistance the European and International Food Industry, considering the applicable Food Law in the widest range (regulatory, competition rules, commercial relationship, custom, international trade, etc.). Furhtermore, he is Legal advisor of *Federalimentare*, Italian federation of all the national associations of the producers of food and beverages, member of *CIAA*; Legal advisor of *Assobirra*, association of the Italian producers of beer and malt, member of Brewers of Europe and secretary of Giury of Self – regulation Code on Beer Advertising; Member of the Executive Committee of *Assocasa*, association of the Italian producers of detergents and soaps; Member of the Legal Affairs Committee of *Federchimica*, federation of the Italian chemical industry, member of *Confindustria*; Member of the Council of EFLA as Vice President; Member of the

Council of *AIDA*, Italian Food Law Association; Member of the *CIBUS* Committee, International Food Fair taking place in Parma every two years; Member of the scientific Committee of the Italian Ministry of Industry and Commerce for the simplification and standardisation of national food and beverages legislation (*Codice di diritto alimentare*), in accordance to the Regulation (EC) No 178/2002; Author of several voices of the legal Data Bank on Food Law '*Diritto alimentare. Mercato e sicurezza*' www.leggiditaliaprofessionale.it; Member of the Scientific Committee of the web legal journal '*Rivista di Diritto alimentare*' and author of several articles on themes of food and beverages law, especially in relation to commercial and fiscal problems; www.rivistadirittoalimentare.it.

Cerkia Barnard

C. Barnard is fellow at the Institute for Food, Nutrition and Well-being (IFNUW) at the University of Pretoria (South Africa).

Giorgio Bocedi

Giorgio Bocedi is a founding partner of GB Avvocati, a young Italian law firm dealing with intellectual property, especially Geographical Indications and trademarks, and food law. After attending a Master of Intellectual Property and Information Society Law at the University of Alicante and several years in two Spanish law firm dealing with intellectual property, currently he is leading the law firm GB Avvocati, established in 2006 in Reggio Emilia (Italy). He handled a number of Geographical Indications cases, concerning the recognition of protection of Geographical Indications within and outside the European union as well as the enforcement of the rights on GIs, including through the Stresa Convention, the Lisbon Agreement on the Protection of Designations of Origin and on International Register and the Council Regulation (EC) No 510/06, on the protection of geographical indications and designations of origin for agricultural products and foodstuff. He has been for several years the legal adviser of the Consorzio del Formaggio Parmigiano-Reggiano and a frequent cconsultant in European and International programs and conferences

Harry Bremmers

B.H.J. Bremmers is Associate Professor of Food Law and Economics at Wageningen University (www.law.wur.nl).

Jorge Cachinero

Jorge Cachinero is Senior Corporate Director of Reputation, Innovation and Corporate Development in LLORENTE & CUENCA and Lecturer at the IE Business School of Madrid, where he obtained his Executive MBA. He is also a member of the Advisory Council of Corporate Excellence – Centre for Reputation Leadership Think-tank. From 1999 until he joined the firm in early 2011, Jorge Cachinero was Director of Corporate and Legal Affairs, Compliance and Communication for the Iberian Peninsula (Spain, Portugal, Andorra and Gibraltar) in Japan Tobacco International. Before that he was Director of Government Relations and Public Affairs in Ford Motor Co. for Spain and Portugal and Secretary of its Board of Directors.

Ricardo Carvajal

Ricardo Carvajal is a Director with Hyman, Phelps & McNamara, P.C., a law firm based in Washington, D.C. that specializes in food and drug law. He provides FDA and FTC regulatory counseling and litigation support to manufacturers and marketers of foods, including dietary supplements and medical foods. He advises clients on a variety of matters, including implementation of the Food Safety Modernization Act; management of Reportable Food Registry incidents, recalls, inspections, and GMP/HACCP enforcement issues; labeling and advertising issues, including those that arise from the use of health, nutrient content, structure/function, and disease claims; regulatory strategy with respect to product development, including requirements pertaining to self-determinations of GRAS status and determinations of new dietary ingredient status; diligence reviews for major acquisitions and initial public offerings; and expert opinions on FDA regulatory matters, including those that arise in the context of criminal prosecution. From 2002 to 2007, Mr. Carvajal served as Associate Chief Counsel in FDA's Office of Chief Counsel, where his work was recognized with several awards. He currently serves on the Editorial Advisory Board for the Food and Drug Law Institute Monographs. Previously, he served as Chair of the Public Policy Outreach and Implementation Task Force of the Institute of Food Technologists (IFT), and as a co-instructor for IFT's annual short course on food labeling requirements in the U.S. He regularly publishes updates on current developments in food law at www.fdalawblog.net. He holds a J.D. from Northwestern University School of Law and an M.S. in Biology from the University of Michigan, and is admitted to practice in the District of Columbia.

Nicola Conte-Salinas

Nicola Conte-Salinas is a lawyer admitted to the Bar in Germany. She specializes in German and European Food Law. She also advises and assists food companies in questions relating to the law against unfair competition, as well as in the area of contract law, in particular as regards general terms and conditions. She has been working in the food law practice of KWG since January 2009. Nicola is the contact person for the Food Lawyers Network Worldwide e. V. (FLN). She is a lecturer for food law at the Hochschule Rhein-Waal (Rhine-Waal University of Applied Sciences), and a regular speaker at food law seminars and conferences. Nicola is the author of various publications on food law issues. Nicola graduated from the University of Trier (Germany) and received a Diploma in Legal Studies (with distinction), as well as a Diploma in Higher Education from the University of Wales, Cardiff (UK). She advises and represents clients in German and English.

Nicole Coutrelis

Nicole Coutrelis is a lawyer at the Paris Bar since 1985. Before starting her private practice, she was at the French 'Office National Interprofessionnel des Céréales' (Cereals Authority) where she was in charge of the European Affairs (1972-1978). She was then nominated at the French Permanent Representation to the EC in Brussels, ('Attaché Agricole' from 1978 to 1982) and joined afterwards the Legal Service of the European Commission (1982-1985). Nicole Coutrelis is a partner of Coutrelis & Assoices, a French Law Firm with Offices in Paris and in Brussels, specializing in EU Law, mainly in Antitrust, Agriculture, Food and Drug regulations, Alcoholic Beverages, Free Movement of Goods and all matters related to the Single Market. She is particularly in charge of the Food Law Practice of the Firm, counseling, lobbying and litigating, in France, in other Member States of the EU and before the European Court of Justice and Court of First Instance in Luxembourg. She is Vice President of the European Food Law Association. Vice President of the International Wine Law Association. She is a Member of the International Bar Association, the Institute of Food technologists and of the Food and Drug Law Institute in Washington. She is the author of many articles and Conferences related to European Food Law.

Mariano López Benítez

Mariano López Benítez is full professor of Administrative Law at the University of Córdoba and is the author of numerous books and articles on Food Law Reviews on administrative law and regulation of Designations or Origin. He has received numerous awards, among them the European Award to the best doctoral thesis on Public Law in 1993.

Angel Martienez Gutierrez

Angel Martínez Gutierrez is full professor of Mercantile Law at the University of Jaen. He studied at the University of Jaen, in Spain, (Law Degree with extraordinary honours) and Bolonia, in Italy, (PHD summa cum laude and Avocatto Mario Jacchia Award). He has received a Mencion de Honor XIII Premio Unicaja on agricultural research Award. Martinez is the author of 5 book, 20 chapters in books and many scientific articles in Law Reviews. He edited the book Marca y Publicidad Comercial (2009) and has been speaker and lecturer in numerous Congress and master degrees in several countries (Italy, Poland, Portugal, Tunisia, Dominican Republic, Mexico, etc.). Angel Martinez produces regularly legal opinions and is responsible of research project in several Universities and companies.

Kai Purnhagen

K.P. Purnhagen is Assistant Professor of Law and Governance at Wageningen University, the Netherlands and distinguished international visitor at the Erasmus University of Rotterdam, Rotterdam Institute of Law and Economics.

Donald Renie

Practised as a lawyer/solicitor from 1972 to 2004, specialising in Agricultural and European law. Formerly president, The European Council for Agricultural Law (CEDR) from 2010 to 2013; Vice Chairman, Scottish Lawyers European Group; External examiner in agricultural law in the University of Edinburgh; Secretary (now Honorary Member) of the Scottish Agricultural Arbiters Association; Member of the Landlord and Tenant Consultative Panel on proposals for reform of the Agricultural Holdings legislation and one of the experts consulted by the Scottish Law Commission in connection with its report into agricultural jurisdictions; Member of the Law Society of Scotland's Working Party on Arbitration; Member of Council of the Agricultural law Association (UK); Member of Scottish Government's Panel of Agricultural Arbiters (until abolition of the Panel); Joint Editor, Connell on the Agricultural Holdings (Scotland) Acts (7th Edition); Contributor to the Encyclopaedia of Scottish Landlord and Tenant Legislation; Author of numerous articles and lectures on topics in rural law, including tenancies, access to the countryside, agricultural quotas, dispute resolution, structures for farming businesses and taxation.

Vicente Rodríguez Fuentes

Vicente Rodríguez is a Spanish lawyer *(abogado)* specialized in food Law. He was educated at the Universities of Seville (law degree) and Liege (Master Degree in European Law). He is the authors of numerous publications on Food Law in Law reviews, in Spain, Italy and Germany. He directed the book *El Derecho Agroalimentario (*Agrifood Law) and participated in the book *UE Sociologia y Derecho Alimentario* (UE Sociology and Food Law). He has also written the legal section of the Spanish Ministry of Health's/FIAB Guidelines on the Food Alert System and a handbook on how to deal with food alerts together with *Llorente y Cuenca* Communication Agency. He has a wide experience of litigation in the field of food law and particularly food alerts in Spain and abroad, having intervened in cases where the alerts were declared illegal by the Spanish Supreme Court. Vicente Rodríguez is a member of Agrofood Legal. Since 2012 he is the president of the European Food Law Association.

Bernd Van der Meulen

B.M.J. van der Meulen (Bernd) was born on Queensday (April 30) 1960. He studied law (1984) and governance sciences (1986) at the Free University (*Vrije Universiteit*) in Amsterdam. At the same university he received his PhD (on enforcement of public order) in 1993. He taught constitutional and administrative law at several Universities in the Netherlands. He practised as an attorney for five years and worked at the Netherlands' Competition Authority (*NMa*) for four years lastly as litigation officer. Currently he is Professor of Law and Governance at Wageningen University. The Law and Governance Group focus their research and teaching on the legal and legal-economic aspects of agro-food-chain (see www.law.wur.nl), including regulatory affairs, intellectual property, comparative food law and human rights.

Introduction

Food is mainly produced by agriculture and agriculture is mainly devoted to the production of food. Food production and agriculture are interconnected and cannot be understood without taking into account that they are both part one of the main integrated economic and cultural activity of human beings: the production, commerce and consumption of food. Inevitably, food law permeates agricultural law and vice versa. Historically, since the creation of the European Economic Community, the regulation of agricultural activities has been one of the main areas of European legislation. In parallel, harmonisation of food law has been essential in achieving a unified European market. While agricultural law was already a well-developed legal discipline, regulated both by the legal traditions of the Member States and by the several treaties that have formed the European Union, food law is a newer field whose limits are not as well defined and whose structure and content are not yet complete. Food law is, in part, consumer law and in part, market regulation. Sometimes it resembles pharmaceutical law and sometimes it is agricultural law.

In 2012, the European Food Law Association devoted its Seville Congress to the study of Food Law and its agricultural dimension. We decided to discuss how food and agricultural law relate to one another and how the boundaries between them are becoming increasingly overlapping. This book is the result of this Congress. The Congress was divided into four different sessions, each with its respective speakers. Similarly, the chapters of the book, written by some of those speakers, reflect the different aspects presented at the conference.

The first session was an introduction in which we analysed how food and agricultural law were evolving. The introduction was carried out by Horacio Gonzalez Alemán, Secretary General of the Spanish Food and Drinks Association (FIAB), who opened the Congress explaining the importance of the production of foods to our economy. Following this introduction, Professor Ferdinando Albisinni described the evolution from agricultural law to food law, from the EEC Treaty to the judicial decisions that have created a corpus of European food law and how these decisions have shaped the current legal system of the European Union. Kai Purnhagen, Cerkia Barnard, Bernd van der Meulen and Harry Bremmers study, in the subsequent chapters, a new dimension of food law - private food law - a concept established during the previous EFLA Congress in Amsterdam, in 2010.

The second session was devoted to the regulation of food quality, which has traditionally been the subject of agricultural law, as applied to designations of origin and geographical indications. This part of the book, by way of continuation to the previous session, begins with a chapter written by Donald Rennie, President of the European Council for Rural Law, on the new Common Agricultural Policy. This is followed by a chapter written by Professor Mariano López on the importance of the regulation of quality as the origin of food law. The regulation of food quality finds its more classic expression in designations of origin and geographical indications. In the following chapter, the lawyer Giorgo Bocedi, sets out how one of the oldest and better-known protected designations of origin, *Parmigiano Reggiano*, defends itself against those attempting to take advantage of the reputation and name of the product. Then Professor Angel Marinez Gutierrez analyses some similarities and differences between the protection of geographical indications and trademarks, in an insightful and critical chapter.

The third section of the Congress was devoted to the control of food-safety and the workings of the food-alert system. Three lawyers from different jurisdictions analyse the situation in their countries, Nicolas Conte-Salina in Germany, Vicente Rodríguez Fuentes (myself) in Europe and Spain and Ricardo Carvajal in the USA. Finally, since the main impact of a food alert affects product reputation, a communications expert, Jorge Cachinero, Group Senior Director at Llorente & Cuenca, addresses in his chapter the importance of managing reputation when affected by a food alert.

The fourth and last session of the Congress was devoted to food trade, to how prices are formed in the market. Nicole Coutrelis contributes to this study with a chapter on the application of competition law to Agricultural Law, both historically and after the reformed CAP. Finally, Alessandro Artom writes an interesting chapter on the regulation of food transactions within the Italian legal system. Moreover, our EFLA colleagues Michel Fogden and Paul Allen assisted us in reviewing the content and language of some chapters of this book. Finally, I have to thank professor Bernd Van der Meulen for all his assistance in the edition of this book.

The content of the presentations and, as a consequence, the content of this book are heterogeneous. Some speakers are academics, others are practising lawyers, some were civil servants while others represent businesses and producers' organisations. They came from different countries and professional backgrounds. This gave the Congress a unique comparative practical/legal perspective and varied professional points of view, which, in my opinion, are very well suited to the hybrid nature of food and agricultural law.

The selection of contributions from the Congress presented here has produced a very interesting book which is a perfect example of the aims of the European Food Law Association, namely, to foster the study of food law. This is another step forward for EFLA, and we aim to continue the initiative in future, with the publication of a new book after every Congress.

I hope you enjoy the discussion, the diversity of points of view and the many experiences contained in this book.

Vicente Rodríguez Fuentes
President of EFLA

1. Agricultural and food law as innovation engine of European law: the new scenario

Ferdinando Albisinni

1.1 Premise

Agriculture Law, Food Law, and Innovation share a peculiar relation, characterized by reciprocal strong influence and interference. Technological innovation, mainly from 19th century, starting with the revolutionary ideas of the famous French chef Appert, has radically changed the techniques applied to food, in all phases from primary production, to processing, to distribution. The same reference to 'tradition' cannot be properly appreciated in the field of agri-food production if not connected to innovation. If, in general terms, we can speak of 'The invention of tradition'[1], or we must admit that 'tradition is a well sorted innovation'[2], these considerations assume peculiar meaning when referred to the agri-food sector.

Quality wines and extra-virgin olive oils – to mention only two renowned flags of 'traditional' Mediterranean food products – have to-day a quality much higher than only a few decades ago, as a result of radical technological innovations, both in the primary phase and in the transformation process.

Food and Agricultural products (even those named as 'traditional') cannot renounce to innovation. But technological innovations – due to their nature of something new, whose effects are far from being fully known and appreciated – offer at the same time benefits and costs (not only in the economic field), advantages and risks (potentially affecting human health or natural environment). In other words, technological innovation affects in a radical way products and processes in the agri-food sector, modifying almost every day what Tullio Ascarelli described as 'the nature of facts'.

As a consequence, Agricultural Law and Food Law are under strong pressure to find adequate regulatory answers to the challenges of technological innovations, in a sort of permanent confrontation between technical innovation exploring new territories, and legal innovation, forced to deal with new issues, new problems, new regulatory areas. In this perspective law is said to react to technological innovation.

[1] Reference is made to the brilliant and well-known oxymoron by Eric J. Hobsbawn, *Come si inventa una tradizione*, in E.J. Hobswan- T. Ranger, *L'invenzione della tradizione*, Cambridge, 1983, It. ed., Turin, 1987.
[2] As stated by C. Barberis, *L'agricoltura tra gastronomia e alimentazione*, in INSOR, *Gastronomia e società*, Milan, 1984, p. 21.

In Europe, the introduction of rules on the traceability of beef after the explosion of the 'mad cow' crisis, the Novel Foods legislation, the general adoption of the precautionary principle, may be cited among the many significant examples of an approach of the food legislator, aimed to answer to needs and demands coming from innovation and not finding sufficient regulatory tools in the already existing legislation.

The impact of technological innovation on the legal framework is not limited to the substantive aspects of food, as it deals largely – in an always more pervasive way – with immaterial aspects, first of all those related to the communication on the market, thus forcing the food legislator to deal both with substantive and immaterial issues in regulating food production and marketing.

But juridical innovation is not only an answer to technology, it is not only reaction. Juridical innovation – with great evidence in the specific area of food law – is also action, proactive innovation. It is by itself an expression of the elaboration of new and original models and institutions, through an experimental approach, which develops in what some German scholars effectively defined as Rechtsreform in Permanenz[3].

As a U.S. scholar observed: 'The most important principles of United States constitutional law have been developed in the context of food legislation in general'[4]. We can therefore conclude that juridical innovation in food law, sharing roots and elements with technological innovation, includes both:

- juridical innovation as reaction; and
- juridical innovation as action.

Such a conclusion appears to be true, more generally, even for European law considered as a whole.

To appreciate the role played by Agricultural and Food Law in developing basic principles of general Community law, it is sufficient to recall the judicial doctrine of mutual recognition, and the large number of decisions of the Court of Justice in the field of food production and food market, defining content and scope of principles of proportionality of administrative action and of protection of European citizen's rights.

Regulation (EC) No 178/2002 not only reaffirmed and enlarged rules already introduced with more limited scope (like traceability or precautionary principle), but moved in the direction of a comprehensive legal system, in which even

[3] U. Noack-D.A. Zetzsche, *Corporate Governance Reform in Germany: The Second Decade*, Center for Business & Corporate Law Research Paper Series No. 0010, Düsseldorf, 2005.

[4] P.B. Hurt, *Food law and policy: an essay*, in *Journal of Food Law & Policy*, 2005, p.1, at p.8.

traditional European rules (such as those referred to labelling marketing, or producers' liability) should be applied in a new integrated context.

In this perspective, interpretation and application of Regulation No 178/2002 and of the large number of European food rules introduced before and after the year 2002, require scholars, legal practitioners, public administrators and business operators, to appreciate the framework within which regulations and directives operate. It is no longer sufficient – as it was until recently – to merely follow individual rules one by one. European Agricultural law and Food Law are building an integrated framework, which includes goals, tools, competences, responsibilities, procedures, and material provisions, emphasizing the role of administrative and judicial experience, and introducing a net structure, within which local, national and European regulators and administrators are called to strict cooperation. One may therefore assume that, as a result of such process, a new legal model is emerging, in which rules coming from EU level and rules coming from national and local levels strictly linked in a unitary model of European law.

In this complex and articulated process, which enriches the armoury of legal tools with new additions, and at the same time enhances consolidation and simplification of existing rules, the traditional borders between public and private law are assuming new contents. Protected interests typically classified as 'public goods' – like consumer's protection and fair competition – increasingly rely on new tools which utilize private models, like contracts, voluntary acceptance of rules and standards, and civil liability as enforcement tool instead of criminal or administrative sanctions. In the field of control and surveillance, certification bodies of private nature are chosen to perform duties which are traditionally considered 'public'.

This chapter will try to highlight the mechanisms through which new models and institutions are emerging, in a proactive and reactive permanent dialogue between innovation and European agriculture and food law.

1.2 The first 50 years of Community Agricultural Law and of Community Food Law

In the year 2012 of the EFLA congress (a year so difficult for European peoples and States) we celebrated a relevant and well defined anniversary: the first 50 years of Community Agricultural Law and of Community Food Law. I think it is proper to situate this 50 years anniversary in 2012, because the first European Agricultural and Food rules having general content and general application have been adopted only in 1962, even if the Common Agricultural Policy was already expressly and separately mentioned in the original text of the EEC Treaty of 1957.

In the first years of existence of the EEC, also after the Stresa Conference of 1958, there have been some acts and decisions adopted assuming CAP as legal basis, but these were intended to deal with single cases and were lacking of general disciplinary content and character[5]. In 1962, by contrast, a number of general provisions have been adopted in a very short period of time. Those general provisions, regarding both Agricultural Law and Food Law, confirm the intimate relationship between these two areas; a relationship which was clearly recognised by the European legislator, even if in the original text of the EEC Treaty no express reference was made to the principles and values, which are nowadays commonly assumed as the basis of the European Food Law[6].

The European Economic Community, already in the first years of its existence, still in its original structure of only six Member States[7], showed a significant awareness of the need to promote the introduction of common rules on foodstuffs, together with the introduction of rules on production and trade of agricultural products, conscious of the quantitative and qualitative relevance of food trade within the market at that time named 'common'[8].

It is necessary to mention here, on the one hand, Regulation (EEC) No 26/62 establishing rules on competition in production and trade of agricultural products[9], and Regulation No 25/62 on the financing of CAP which created the European Agricultural Guidance and Guarantee Fund[10], both of April 20, 1962, and on the other hand a series of Regulations, No 19/62[11], 20/62[12], 21/62[13], 22/62[14], 23/62[15], 24/62[16], adopted in the same day, introducing the first CMOs, on cereals, pigmeat, eggs, poultrymeat, on fruit and vegetables, and on wine.

[5] For an historical analysis of the first years of the CAP, which underlines the novelties introduced in 1962, see L. Costato, *Agricoltura*, in *Trattato di diritto amministrativo europeo*, M.P. Chiti and G. Greco (eds.), Milano, 1997, I, p. 1 ff.
[6] See L. Costato, *Principles and rules of European Food Law*, in L. Costato and F. Albisinni (eds.), *European Food Law*, Padova, 2012, cap. I.
[7] See EEC Treaty 1957, establishing the European Economic Community.
[8] Article 3 TEEC 1957.
[9] EEC Council Regulation No 26, of 20 April 1962, applying certain rules of competition to production of and trade in agricultural products.
[10] EEC Council Regulation No 25, of 20 April 1962, on the financing of the common agricultural policy.
[11] EEC Council Regulation No 19, of 20 April 1962, on the progressive establishment of a common organisation of the market in cereals.
[12] EEC Council Regulation No 20, of 20 April 1962, on the progressive establishment of a common organisation of the market in pigmeat.
[13] EEC Council Regulation No 21, of 20 April 1962, on the progressive establishment of a common organisation of the market in eggs.
[14] EEC Council Regulation No 22, of 20 April 1962, on the progressive establishment of a common organisation of the market in poultrymeat.
[15] EEC Council Regulation No 23, of 20 April 1962, on the progressive establishment of a common organisation of the market in fruit and vegetables.
[16] EEC Council Regulation No 24, of 20 April 1962, on the progressive establishment of a common organisation of the market in wine.

In those first Regulations, adopted assuming CAP as legal basis but introducing rules in both fields of Agricultural and Food Law, we can find an element, which will be a constant characteristic of general European legislation: the intimate crossing intersection between law and economy, regulation and financing, regulating provisions and financing provisions.

As regards the institutional framework, those regulations introduced procedures, competences, offices, boards, which during the years became common to the entire European system. It is sufficient here to mention the introduction of the Committees by the first CMOs[17], which gave origin to the s.c. 'comitology' [18] and it is still one of the peculiar characteristics of the European constitutional architecture[19].

1.3 Common market and health protection in the first steps of approximation

The evolutionary process towards the establishment of a unitary common framework of Agricultural and Food Law in all its aspects, related to production, trade, marketing, health and safety, was slower than could be anticipated in the first years of the EEC and found its way through the contribution of a multiplicity of factors and sources.

It is possible to outline some temporal partitions, which (with all the arbitrariness and tentativeness of any partition) characterise this process, in significant correspondence with institutional developments, both in the internal European dimension, and in the international dimension of growing globalisation, together with the challenges of technological and market innovations.

In 1962, a few months after the Regulations of April 1962, mentioned above, the European Council intervened specifically on food matters, adopting Directive 62/2645/EEC of 1962[20], to regulate colouring matters authorised for use in foodstuffs intended for human consumption. This Directive was adopted on the basis of

[17] See Articles. 12, 13 of Regulation No 23/1962 on fruit and vegetables, and Articles 6, 7, 8 of Regulation No 24/64 on wines.

[18] See L. Costato, *Agricoltura*, cit., p.24 ff.; M. Savino, *I comitati dell'Unione europea. La collegialità amministrativa negli ordinamenti compositi*, Milano, 2005; D. Bianchi, *La politique agricole commune (PAC) – Toute la PAC, rien d'autre que la PAC ! Précis de droit agricole européen*, Bruylant, Bruxelles, 2012, II ed.

[19] See H.J. Blanke – S. Mangiameli (eds.), *The European Union after Lisbon. Constitutional Basis, Economic Order and External Action*, Springer. Berlin Heidelberg, 2012; P. Bilancia, *The Dynamics of the EU Integration and the Impact of the National Constitutional Law – The European Union after the Lisbon Treaties*, Giuffré, Milano, 2012.

[20] Council Directive of 23 October 1962, on the approximation of the rules of the Member States concerning the colouring matters authorised for use in foodstuffs intended for human consumption (62/2645/EEC).

Article 100 TEEC[21], and therefore on the legal basis of the competence awarded to the Council, 'acting unanimously on a proposal from the Commission [to] issue directives for the approximation of such provisions laid down by law, regulation or administrative action in Member States as directly affect the establishment or functioning of the common market'.

'Establishment or functioning of the common market' were the key words to justify this early EEC legislative initiative, intended to approximate national food laws (i.e. in fact, to unify and render uniform), as expressly pointed out:

> Whereas differences between national rules concerning these colouring matters hinder the free movement of foodstuffs and may create conditions of unfair competition, thereby directly affecting the establishment or functioning of the common market[22].

But the Directive, even before mentioning 'free movement' and 'common market' in Recital 2, in Recital 1 opens the text with the basic consideration that 'all rules relating to colouring matters which may be used in foodstuffs intended for human consumption must give priority to the protection of public health'[23]. 'Protection of the consumer against falsification', together with the 'needs of economy' are expressly mentioned in the first Recital of the Directive, as reasons that 'must also be taken into consideration'.

One year later, on the same legal basis of Article 100 TEEC, Directive 64/54/EEC[24] introduced provisions for the approximation of national legislation on preservatives used in foodstuffs intended for human consumption, with the even clearer specification that the value of 'health protection' is so prominent, that even 'economic and technological needs' can be taken in consideration only 'so far as health protection allows'[25]. Health protection is therefore identified – even at language level – as such a prominent value, that it cannot even be weighted in comparison with other interests already mentioned in the Treaty, but must in any case prevail, due to the highest value attributed to human beings. In other words, no balance of interests or cost-benefit analysis is envisaged in the Directive of 1964; instead an express hierarchy of values and interests is established, and health protection is situated at the top.

From these first steps, European legislation on food revealed its pluralistic nature, given its multiple objectives, and at the same time affirmed a founding rule of

[21] Later Article 94 TEC, and now Article 115 TFEU.
[22] Recital 2 of Directive 62/2645; italics added.
[23] Recital 1 of Directive 62/2645; italics added.
[24] Council Directive 64/54/EEC of 5 November 1963, on the approximation of the laws of the Member States concerning the preservatives authorized for use in foodstuffs intended for human consumption.
[25] Recital 1 of Directive 64/54/EEC; italics added.

action, which, in a Community at that time still named Economic Community, privileged health protection on any protected interest, including economic interests. The expression food safety at that time was not in use in the regulatory language of the EEC. Even the expression food security was not commonly used. But the substance of food security was mentioned among the express objectives of the Common Agriculture Policy (CAP), which included those 'to assure the availability of supplies' and 'to ensure that supplies reach consumers at reasonable prices'[26].

In coherence with the integrated approach adopted by Articles 38-47 TEEC and the multiple objectives assigned[27], the CAP in many ways characterised itself as a complex agro-food policy, more than a mere agricultural policy, dealing also with phases subsequent to the primary activity, including transformation, distribution, trade and marketing, in a framework which was very different from many national legal systems (e.g. Italy, where the area of agricultural law was traditionally limited to the primary activity *per se*, enclosed within the borders of the farm).

It is sufficient here to mention Regulation No 136/66[28] on oils and fats, adopted on the basis of Articles 42-43 TEEC, which in 1966 adopted common customs tariffs, established a system of compensatory amounts, import export licences, market target prices, intervention prices and threshold prices (with complex administrative procedure and bodies to apply the intervention rules), but at the same time introduced uniform definitions, including those of virgin olive oil, refined olive oil, pure olive oil[29], directly addressing aspects which in national legislations were traditionally considered part of food law.

With an analogue approach, Regulation No 804/68[30] on milk and milk products, also adopted on the basis of Articles 42-43 TEEC, in 1968 regulated the target price for milk and the intervention price for butter and cheese (with specific intervention prices for Grana Padano and Parmigiano Reggiano)[31], adopted common customs tariffs, import export licences, export refunds (also in this case with complex administrative procedures and bodies to apply the intervention rules), but at the same time introduced uniform quality standards applicable to 'Community-produced butter' definitions, establishing that only butter which reaches such quality standards may be imported into the EEC territory[32].

---

[26] Article 39.1.(d) TEEC in its original text (now Article 39 TFEU).
[27] See now Artt. 38-44 TFEU.
[28] Regulation No 136/66/EEC of the Council of 22 September 1966, on the establishment of a common organisation of the market in oils and fats.
[29] Annex to Regulation No 136/66/EEC.
[30] Regulation (EEC) No 804/68 of the Council of 27 June 1968, on the common organisation of the market in milk and milk products.
[31] Article 5.1. Regulation No 804/68.
[32] Article 16.1. Regulation No 804/68.

1.4 The '70s: partial approximation and CAP measures

The '70s, with the enlargement of EEC to nine members, consequently including the United Kingdom, Ireland and Denmark, countries with significantly different agriculture and food traditions from countries like France and Italy, made the need to proceed to an approximation of rules even more pressing.

Notwithstanding Council Resolution of 28 May 1969[33] aiming to eliminate technical obstacles to the exchange of foodstuffs, and the more general Directive 50/70/EEC of the Commission on the abolition of measures which have an effect equivalent to quantitative restrictions on imports[34], the approximation of technical rules regarding food products remained an unfinished process for a long time.

Exemplary is the case of chocolate. A Council Directive of 1973[35], adopted on the legal basis of Article 100 TEEC, and therefore by unanimous consent of all MS, admitted that, due to the great differences among national legislations and traditions, at that time it was not possible to completely harmonise all the provisions relating to these products and lay down common definitions and composition rules, and therefore expressly postponed effective harmonisation to a later date[36]. This later date arrived only three decades later, with a Directive of 2000[37].

As a consequence of the impasse in the harmonisation process, the recourse to CAP measures under Article 43 TEEC, already used in the '60s to regulate the characteristics of food products (as in the aforementioned cases of oils and fats with Regulation No 136/66, and of butter with Regulation No 804/68), became the *de facto* standard legislative tool utilised during the '70s and large part of the '80s, to adopt rules in areas of food law.

Many provisions adopted within the CAP dealt with the Common Organisation of agricultural Markets (CMO)[38]. Those acts, adopted by the European Council exercising the powers and competences assigned by Article 43 TCEE and under the procedure thereby established, were intended to guarantee citizens the availability of supplies (i.e.: food security), farmers increased incomes and a fair standard

[33] Council Resolution of 28 May 1969, drawing up a programme for the elimination of technical barriers to trade in foodstuffs which result from disparities between the provisions laid down by Law, Regulation or Administrative Action in Member States.
[34] Commission Directive 70/50/EEC of 22 December 1969.
[35] Council Directive 73/241/EEC of 24 July 1973, on the approximation of the laws of the Member States relating to cocoa and chocolate products intended for human consumption.
[36] Recital 5, 7, 8, of Directive 73/241/EEC.
[37] Directive 2000/36/EC of the European Parliament and of the Council of 23 June 2000, relating to cocoa and chocolate products intended for human consumption.
[38] Article 40 TEEC in its original text.

of living, and consumers reasonable prices, in accordance with the objectives of CAP established by Article 39 TCEE[39].

As in the '60s, those acts, together with economic content, fixing prices and establishing financial aid to farmers, frequently went on regulating food products, within the framework of a regulated market, in some cases with reference to food products situated well beyond the category of 'products of first-stage processing directly related to these products' mentioned in Article 38 TEEC.

In this context, the expansive capacity of CAP regulation is confirmed by the circumstance that most of the Directives expressly aimed to pursue harmonisation in the field of foodstuffs and of feeding-stuffs have been approved on the double legal basis of Articles 43 and 100 TEEC, considered jointly.

In other words, during the '70s and in subsequent years, CAP measures operated to promote an approximation (at least partial) that the provisions of the Treaty expressly dedicated to harmonisation were not able to assure, and Article 43 TEEC on CAP played a decisive role in the rule-making process of the food law discipline, offering some sort of founding pillar.

This regulatory process led to the adoption both of vertical rules (harmonisation measures for single food chains and products), and of horizontal rules (regarding the generality or at least large aggregates of products).

In the first group, the list includes some Directives approved in the '70s, on the approximation of laws regarding single food products: e.g. honey[40], fruit juices[41], preserved milk for human consumption[42], fruit jams, jellies and marmalades and chestnut purée[43], all adopted on the double legal basis of Article 43 and 100 TEEC considered jointly.

In the second group, the list includes some Directives, also adopted on the double legal basis of Article 43 and 100 TEEC considered jointly, regarding decisive issues of food safety and health protection (which nowadays are regulated by the 'Hygiene package', as a relevant part of the present food law system), among which:

[39] Reaffirmed unchanged in Article 39 TFUE, as introduced by the Lisbon Treaty, in force starting from 1 December 2009.
[40] Council Directive 74/409/EEC of 22 July 1974.
[41] Council Directive 75/726/EEC of 17 November 1975.
[42] Council Directive 76/118/EEC of 18 December 1975.
[43] Council Directive 79/693/EEC of 24 July 1979.

- two Directives of 1970, one on the methods of sampling and analysis for the official control of feeding-stuffs[44], and the other on additives in feeding-stuffs[45], and
- two Directives of 1976, one relating to the fixing of the maximum level of erucic acid in oils and fats intended for human consumption[46], and the other relating to the fixing of maximum level for pesticide residues in and on fruit and vegetables[47].

In all those cases, measures adopted within the CAP significantly intervened in the food market, introducing some basic elements of a food law discipline rooted in agricultural law; a circumstance which largely explains the growing interest of agricultural law scholars in themes of food law.

Only a few acts, nearing the end of the '70s, were approved on the basis of Article 100 alone, *per se*, among which:

- a Directive of 1976, on materials and articles intended to come in contact with foodstuffs[48], and not on food products as such;
- a Directive of 1977[49], on coffee extracts and chicory extracts, which adopted the same approach already adopted by the mentioned Directive of 1973 on cocoa and chocolate, leaving to national provisions a large possibility to maintain derogations to the harmonized provisions, with reference both to the characteristics of the products, and to names and labels, confirming a situation of substantial stall, with a partial and unfinished harmonisation.

The important Directive on labelling, presentation and advertising[50] must be added to this short list. This Directive, adopted at the end of the '70s on the basis of Article 100 TEEC, introduced horizontal rules applicable to the generality of foodstuffs to be delivered to the ultimate consumer. With this Directive, the process of harmonisation goes a step further in the area of market communication, but does not enter into the merits of production processes, and does not deal with the sensitive issue of the 'names' of the foodstuffs, accepting a persistent non-harmonization[51].

---

[44] Council Directive 70/373/EEC of 20 July 1970.

[45] Council Directive 70/524/EEC of 23 November 1970.

[46] Council Directive 76/621/EEC of 20 July 1976.

[47] Council Directive 76/895/EEC of 23 November 1976.

[48] Council Directive 76/893/EEC of 23 November 1976.

[49] Council Directive 77/436/EEC of 27 June 1977.

[50] Council Directive 79/112/EEC of 18 December 1978, on the approximation of the laws of the Member States relating to the labelling, presentation and advertising of foodstuffs for sale to the ultimate consumer.

[51] See Article 5(1) Directive 79/112/EEC – original text.

1.5 The '80s: judicial interventionism

Even in the '80s, others provisions regarding food law were introduced by CAP measures on the Common Organisation of agricultural Markets (CMO). But the material shift toward a European Food Law (EFL) system, with common shared principles, characterised by a plurality of values, interests and objectives, and by a complex multi-level rule-making system, took place on a new original bases, in strict relation with the critical issues (both of food safety, and of economic development and competition) raised by the removal of national borders and customs and by the process toward a unified European domestic dimension.

During the '80s, the accession to the EEC of three Mediterranean countries, Greece, Spain and Portugal, all with relevant and specific agro-food productions, deeply rooted in their respective traditions and a significant part of their economic systems, gave rise to an enrichment of diversities. As a result, the need to remove national obstacles to free trade of foodstuffs became more urgent, requiring innovative legal tools.

The new pillars of legal innovation, which contributed to the building of the EFL system, may be identified in:

- the construction and consolidation of the judicial doctrine of 'mutual recognition' through the creative work of the Court of Justice;
- the construction of the internal market, through the engagement of the European Institutions after the approval of the Single European Act in 1986 and of the Maastricht Treaty in 1992.

The doctrine of 'mutual recognition' is the decisive contribution of the Court of Justice, in answer to the substantial stalemate position in the regulatory strategy intended to assure free circulation of goods in the European Community.

The judicial unification of the European market moved from a broad interpretation of the prohibition of 'measures having an equivalente effect' to quantitative restrictions on imports and exports, established by Articles 31 and 34 TEEC.

The case which opened the way was the well-known *Dassonville* case of 1974[52]. Starting from the rationale of this decision and confirming the broad interpretation of Articles 31 and 34 TEEC, in 1979 the leading case *Cassis de Dijon*[53] introduced the principle of 'mutual recognition', adopting the 'equivalence' principle as the basis to unify the internal market while respecting diversities, and effectively opening national markets, obtaining through equivalence the result that previously could not be obtained through legislative harmonization.

[52] ECJ, 11 July 1974, Case 8/74, *Dassonville*.

[53] ECJ, 20 February 1979, Case 120/78, *Rewe Zentral*.

During the '80s, mutual recognition, initially applied to objective characteristics of food products, was progressively extended to the use of names. The Court of Justice reached the conclusion that a food product, sold in one MS with a certain name, must be admitted for sale with the same name in any other MS, even when it does not have the characteristics and qualities required by the national law of the MS where it is imported, save the introduction of supplementary indications and information on the label[54].

1.6 The '90s: food safety as founding pillar of the internal market

The second pillar of the EFL system emerged in the '90s, within the construction process of the internal market, started with the Single European Act of 1986, consolidated with the Maastricht Treaty of 1992, and required by the enlargement of the EU to new MS[55] and by the need to tackle the challenges of globalisation and of the new international order introduced by the Marrakech Treaty of 1994 and by the subsequent negotiations.

The regulatory scenario changed radically with the introduction in the TEEC of Article 100a[56], on the adoption of 'the measures for the approximation of the provisions laid down by law, regulation or administrative action in Member States which have as their object the establishment and functioning of the internal market'[57], which assigned the regulatory competence to the majority of Council (together with the EP, under the cooperation procedure and later under the co-decision procedure), instead of assigning it to the unanimity as requested by the original Article 100 TEEC (later Article 94 TEC).

The new provision permitted to extend the Community rule-making process to many sectors of food production and market, previously regulated only by national rules. Article 100a was in fact largely used to overcome regulation differences among MS; differences which were originally much greater for food products than for other products, by reason of the strong territorial roots of food traditions as an element of identity and cultural heritage.

As to the content of the foreseen new measures, Article 100a specified: 'The Commission, in its proposals envisaged in Paragraph 1 concerning health, safety, environmental protection and consumer protection, will take as a base a high level

[54] ECJ, 14 July 1988, Case 90/86, *Zoni*; 12 March 1987, Case 178/84, *Commission of the European Communities v Federal republic of Germany*; 26 June 1980, Case 788/79, *Gilli*.
[55] Austria, Finland and Sweden.
[56] Introduced by the Single European Act, and partly modified by the Maastricht Treaty.
[57] Article 100a as amended by the Maastricht Treaty; Later Article 95 TCE, and now Article 114 TFUE, with a drafting partially different, which leaves the special legislative procedure and adopts the ordinary legislative procedure.

of protection'[58]. Such provision was in harmony with the rules, also introduced by the Single European Act and the Maastricht Treaty, extending the competence of the Community to the protection of public health (Article 129)[59], to consumer protection (Article 129a)[60], and to preserving, protecting and improving the quality of the environment and assuring a rational utilization of natural resources (Article 130r).

The objectives of obtaining a high level of protection of health, together with the establishment of the internal market, are largely present in food legislation in those years, adopted on the basis of Article 100a.

A relevant step on the path towards the EFL system, as a legal system common and shared among Member States, may be seen in two Council Directives of June 1989, which in an innovative and explicit way expressed a renewed interest in the theme of hygienic control of food products: Directive 89/396/CEE[61] and Directive 89/397/CEE[62].

The first Directive introduced provisions on the identification of the 'lot' to which a foodstuff belongs[63], adopting a definition of 'lot' of general application, according to which 'lot means a batch of sales units a foodstuff produced, manufactured or packaged under practically the same conditions'[64].

The second Directive introduced general principles at European level, 'for the performance of official control of foodstuffs'[65].

The legal basis of both Directives is Article 100a TEEC, i.e. the above mentioned provision, introduced by the Single European Act of 1986, which entered in force only two years before the two Directives, and aimed to favour the establishment and functioning of the internal market.

It must be underlined that Article 2(2) of Directive 89/396/EEC exempted from the application of the new rules the agricultural products which, on leaving the holding, are 'sold or delivered to temporary storage, preparation or packaging stations; transported to producers' organizations; or collected for immediate integration into an operational preparation or processing system;'. Such provision confirmed the

[58] Article 100a(3) TEEC; italics added.
[59] Later Article 152 TEC, amended with specific reference to cover 'measures in the veterinary and phytosanitary fields which have as their direct objective the protection of public health'.
[60] Later Article 153 TEC.
[61] Council Directive 89/396/EEC of 14 June 1989, on indications or marks identifying the lot to which a foodstuff belongs.
[62] Council Directive 89/397/EEC of 14 June 1989, on the official control of foodstuffs.
[63] Article 1(1) of Directive 89/397.
[64] Article 1(2) of Directive 89/397.
[65] Article 1(1) of Directive 89/397.

prevailing model which, until the mad cow crisis and the approval of Regulation No 820/1997 and later of Regulation No 178/2002, focused on food only after the primary agricultural phase, exempting this phase from rules ordinarily applied to the manufacturing and transformation phases of the food chain.

Even with this limitation, adopted for reasons largely related to political and social considerations, the rules introduced by this Directive manifested their nature as elements of a general framework legislation, aimed to offer a systemic design, within which further provisions could be progressively inserted, at National and European level.

A similar approach is present in Directive 89/397/EEC on the official control of foodstuffs. The objectives pursued are more complex than those traditionally assigned to national rules on foodstuff, and cover broader regulation areas than those typical of long lasting national criminal legislations on the composition and the objective hygienic quality of foodstuff.

A further decisive step in the construction of the EFL system was marked in 1993 by Directive 93/43/EEC on the hygiene of foodstuff[66]. Also this Directive indicated Article 100a TCEE as its legal basis and, only a few months after the entry into force of the Maastricht Treaty (1 January 1993), adopted an integrated approach. Obtaining confidence in the standards of safety of foodstuff, guaranteeing the respect of standards throughout all stages of the food chain, counting on the self-responsibility of food business operators, are the key words in this Directive, which introduced the HACCP method in the legal armoury of rules on production and trade of foodstuff, borrowing this method from the space industry, and bringing relevant innovations to the regulatory framework.

The Directive adopted a unitary category of 'food business' covering all stages after primary production and including 'any undertaking, whether for profit or not and whether public or private, carrying out any or all of the following: preparation, processing, manufacturing, packaging, storing, transportation, distribution, handling or offering for sale or supply of foodstuffs'[67].

The agricultural phase of food production remained exempted from the application of the new rules and responsibilities, but all other stages along the food chain were submitted to a unitary regime.

Food operators are called to play a proactive role in the pursuit of food safety, requiring them to invest in education, administration and management (with possible consequent difficulties for small traditional producers, not accustomed

[66] Council Directive 93/43/EEC of 14 June 1993, on the hygiene of foodstuffs.

[67] Article 2 Directive 93/43; italics added.

to and not prepared for such organisation), but at the same time behaviour and production methods acquire much greater relevance than is usual in national provisions.

The rules establishing 'what to do' and fixing 'product characteristics and qualities' are integrated, with Directive 43/93, by rules on 'how to do'. The result - only seemingly paradoxical - is that the Directive of 1993, aimed at the completion of the internal market, adopted a model of harmonization, based on self-responsibility, which has given single food operators ample opportunities of differentiation and variety, as long as the criteria and goals of HACCP are accomplished. It is therefore possible to see the emerging tendency toward the construction of a European Food Law (EFL) system, which in a short time span starts to be provided with rules of general application, together with provisions of specific interest.

1.7 Food quality and food diversity in agricultural legislation

Together with the mentioned objectives prompting the completion of the internal market through 'confidence in the standard of safety of foodstuff'[68] and 'protection of human health'[69], other interests and values emerged as worthy of legislative attention, above all issues of food quality and food diversity.

Those issues found innovative regulation through the adoption of new provisions of agricultural law, which, moving from the primary production, expanded their field of action to regulate market communication and competition.

The judicial doctrine of 'mutual recognition' had removed national barriers to the free circulation of foodstuffs, but had led to the risk of 'banalisation' of traditional productions, neglecting territorial and cultural roots.

In answer to those concerns, Regulations No 2081/92 on PDOs and PGIs[70] and No 2082/92 on AS[71], adopted on the basis of Article 43 TEEC, and therefore as part of the CAP, introduced a new unified regulatory frame, applicable to a large number of quality productions, including not only agricultural products defined as such by Annex II TEEC, but also other agricultural products and - significantly - foodstuffs not included in the list of Annex II and mentioned in further lists expressly introduced by the new regulations[72].

---

[68] Recital 1 of Directive 93/43; italics added.
[69] Recital 2 of Directive 93/43; italics added.
[70] Council Regulation (EEC) No 2081/92 of 14 July 1992, on the protection of geographical indications and designations of origin for agricultural products and foodstuffs.
[71] Council Regulation (EEC) No 2082/92 of 14 July 1992, on certificates of specific character for agricultural products and foodstuffs.
[72] Article 1 Regulation No 2081/92, and Article 1 Regulation No 2082/92.

With the adoption of those acts, the European agricultural rules were expanded to include new territories, in a proactive approach, not limited to financing provisions, but enlarged to include the creation of new goods (the geographical indications), defined at European level and taking into account the international negotiations which two years later, in 1994, would lead to the Marrakesh agreement setting up the WTO and to the TRIPs agreement.

The emerging scenario, at the beginning of the '90s, is that of a pluralistic dimension of food law, still missing the comprehensive general law which later would be introduced by Regulation No 178/2002, but largely rooted in agricultural law and characterised by the European attention to peculiar qualities (and to cultural and territorial roots) of foodstuffs, recognizing the peculiarities of production methods and products[73] and the related need of diversified disciplines, referring both to products and production methods as well as to labelling and market communication.

In the same perspective, Regulation No 2092/91 on organic production of agricultural products[74], also adopted in the early '90s on the basis of Article 43 TEEC as part of the CAP, considered both the primary phase of production and the following phases of processing, preserving and packaging, and introduced provisions which regulated the production method and the inspection system, together with labelling and marketing, adopting a unitary regulatory approach, applicable to the entire food chain.

1.8 Food safety crises and the new disciplinary framework

The reference frame was dramatically modified in the late '90s, due to some serious food safety crises, which gained large repercussion amongst the general public: from BSE (bovine spongiform encephalopathy) to dioxin.

As a result, general confidence and trust in the effectiveness of existing rules on food safety was largely undermined, the market in beef and beef products was heavily destabilized, and the European legislator was pressed to find rapid answers, within an approach which, by its nature and its object, required an overall systematic design.

The regulatory solutions adopted were situated at different levels and operated through different tools. On a technical level, some specific hygienic measures were introduced, e.g. the prohibition to use meal of animal origin to feed bovine animals, the prohibition to trade and use some parts of bovine animals, and the duty to slaughter the entire cattle herd in case of suspected BSE. But the most

[73] Recital 3 Regulation No 2081/92.

[74] Council Regulation (EEC) No 2092/91 of 23 June 1991, on organic production of agricultural products and indications referring thereto on agricultural products and foodstuffs.

original answer was introduced on the administrative level by Regulation No 820/97[75], with the adoption of a new legal tool, until then unknown to the legal jargon (except limited cases in finance and public procurement law) and common only in scientific language: traceability.

Traceability may not be defined as a hygienic measure as such. It refers to a series of provisions including identification, controls, and documentation, which focuses on self-responsibility asking food business operators to adopt a proactive participative role in food law (as happened with Directive 93/43/CEE and the HACCP method), and introduces basic elements to favour the dialogue between controllers and producers.

The entire path of beef products is documented, using specific tools, such as the 'animal passports' which each bovine animal should receive a few days after its birth[76]. The new Regulation for the first time fully involved the primary production in the responsibilities for food safety of beef products, strongly innovating the disciplinary framework of the food chain. With these legal innovations the European legislator reacted in an innovative and original way to the challenges posed by technical innovations in the bovine food sector, not limiting its intervention to the consolidated area of hygienic rules, but moving toward the adoption of provisions strongly innovative in their legal design.

The measures adopted included two groups of provisions:

- those introducing a system for the identification and registration of bovine animals, with the provision of a reference code ensuring the link between, on the one hand, 'the identification of the carcase, quarter or pieces of meat' and, on the other hand, the individual animal[77]; i.e. ensuring traceability;
- those introducing compulsory *labelling from large areas*, establishing that the label on any individual piece or pieces of meat or on their packaging material should include indications of:
  - Member State, third country or holding of birth;
  - Member States, third countries or holdings where all or any part of fattening took place, with partial fattening having to be specified;
  - Member State, third country or slaughterhouse where slaughter took place[78].

Together with the content, the interest of the new regulation lies in the legal basis adopted by the Council, Article 43 TEC alone and therefore CAP. The Commission brought an action under Article 173 TEC against Regulation No 820/97, contesting not the content of the rules, but the legal basis adopted, assuming that it should

[75] Council Regulation (EC) No 820/97 of 21 April 1997, establishing a system for the identification and registration of bovine animals and regarding the labelling of beef and beef products.

[76] Article 6 Regulation No 820/97.

[77] Articles 14 and 16 Regulation No 820/97.

[78] Article 16 Regulation No 820/97.

be approved on the basis of Article 100a (and therefore with the co-decision of the European Parliament), and not on the basis of Article 43 (which left the adoption to the majority of the Council). The European Parliament supported the action of the Commission.

The Court of Justice dismissed the application, with a decision of 4 April 2000[79], which declared legitimate the recourse to the CAP legislative procedure, assuming that marketing rules and restoring consumers' trust are a proper component of the CAP.

In this context, the Court rejected the distinction between production and marketing, and between rules to be respected by producers and rules addressed to consumers: it thus declared the appeal to be unfounded, with the following exemplary explanatory statement:

> ... Article 43 of the Treaty is the appropriate legal basis for any legislation concerning the production and marketing of agricultural products listed in Annex II to the Treaty which contributes to the attainment of one or more of the objectives of the common agricultural policy set out in Article 39 of the Treaty. ... As regards the aim of the contested regulation, it must be observed that, according to the first recital, it is intended to re-establish stability in the beef and beef products market, destabilised by the BSE crisis, by improving the transparency of the conditions for the production and marketing of the products concerned, particularly as regards traceability. ... It must therefore be held that, in regulating the conditions for the production and marketing of beef and beef products with a view to improving the transparency of those conditions, the contested regulation is essentially intended to attain the objectives of Article 39 of the Treaty, in particular the stabilisation of the market. It was, therefore, rightly adopted on the basis of Article 43 of the Treaty.[80].

With this argument, the Court of Justice recognised to Regulation No 820/97, more than to any other previous regulatory act, the nature of exemplary act, looking to agriculture and food legislation within an integrated systemic framework, unifying in a plurifunctional discipline market transparency, competition issues, health protection goals, getting over the borders among different fields of law, and joining in a unified shared legal ambit the consumers and all actors of the food chain, including those operating in the primary phase, therefore stressing the intimate relation between agricultural and food law.

---

[79] ECJ, 4 April 2000, Case C-269/97, *Commission of the European Communities v Council of the European Union.*

[80] See Paragraphs 47, 53, 59, 60 of the Judgement.

This underscored the impetus and the timing of the rise to prominence of a model that was remarkably innovative as compared to the past.

In the new model, the agricultural phase of production, the food processing phase and that of marketing and consumption were combined in an original and avowedly innovative and unified regime.

During those same years, with the *Rioja* verdict, which also dated from 2000([81]), the Court of Justice, prompted by the conflict that has arisen between Belgium and Spain, declared the legitimacy of the Spanish rules that required bottling to be carried out in the area of origin of the Q.W.P.S.R. (quality wines produced in specified regions). The arguments put forward and endorsed in this verdict are particularly significant. First and foremost, it is of interest to mention the argument put forward in the opening statement of the grounds for the sentence, where it is asserted that there is a need to take note of the fact that:

> Community legislation displays a general tendency to enhance the quality of products within the framework of the common agricultural policy, in order to promote the reputation of those products through, inter alia, the use of designations of origin which enjoy special protection. That general tendency has become apparent in the quality wines sector[82].

This statement indicates that agricultural policy was taken as the canon governing marketing as well as production. The Court then underlined that the responsibility for the good reputation of a product with a designated origin lies fully and collectively with the community of producers of the area involved, in the sense that they who are in the best position to guarantee an effective system of quality control.

What emerged is an idea of the *terroir* as a canon of identity and guarantee: that is to say, not a mere reference to a geographic fact, but instead an identification with the locally established community of producers, farmers and non-farmers (including the bottling firms operating in the region). In terms of the market, this canon emphasises the value of the sense of belonging, attributing its benefits to a collective community considered as a whole, and – with the authoritativeness of the Court of Justice – the thus officialised concept of *terroir*, rooted in agricultural law and extended to food law, responds to the demand for a declared link between the policy of upholding the quality of agricultural productions, the rootedness of the population in a given rural area, and the enhanced value of the signs of belonging whit whom the product is presented on the global market.

---

[81] Judgment of the Court of 16 May 2000, Case C-388/95 *Kingdom of Belgium v Kingdom of Spain.*

[82] See Paragraph 53 of the Judgement.

1.9 The 21st century – Regulation No 178/2002 and the systemic dimension

The declared step toward the achievement of a systemic dimension in food law, at the end of the 20th century and in the first years of the new century, is marked by the White Paper of 2000 of the European Commission on Food Safety[83], which planned as many as 84 different administrative and regulatory actions and measures, intended to design an overall framework.

The characteristic element of the entire proposal was the integrated food chain approach, with the inclusion of the agricultural phase of production into a general project of food safety, intending to cover 'all aspects of food products from "farm to table"', as indicated by the Commission with an expression which soon would become famous.

The expression well illustrates the global perspective of food safety, spreading responsibility over all the actors of the food chain, including farmers until then only partially and exceptionally subject to general rules, save the anticipatory case of Regulation No 820/97 on bovine animals.

The architrave of the systemic plan was rapidly implemented with the approval of Regulation No 178/2002[84], whose Chapter II, named 'General Food Law', makes specific reference to the 'principles laid down in Articles 5 to 10 ... [to] form a general framework of a horizontal nature'[85].

The preamble of the Regulation indicates as legal bases, in one single context, Articles 37 (Common Agriculture Policy), 95 (Approximation of national provisions), 133 (Common commercial policy), 152(4)(b) (Health protection), thereby evidencing - even at formal level - the multiplicity of objects, values, interests and goals affected by food law legislation.

In other words, the multiplicity of legal bases corresponds to the plurality of trasversal objects and goals, dealing with multiple areas and needs, and to the innovative character of Regulation No 178/2002, adopting new or newly designed legal models and tools. Traditional borders between production and communication rules are weakened, and the prevalent regulatory criteria is that of responsibilities, both public and private, with a functional design of governance[86].

---

[83] Commission of the European Communities, White Paper on Food Safety, Brussels 12 January 2000, COM(1999) 719 final.
[84] Regulation (EC) No 178/2002 of the European Parliament and of the Council of 28 January 2002, laying down the general principles and requirements of food law, establishing the European Food Safety Authority and laying down procedures in matters of food safety.
[85] Article 4(2) Regulation No 178/2002.
[86] See Article 17, entitled '*Responsibilities*' of Regulation No 178/2002.

Undertakings and business operators are called to operate within an integrated framework of regulatory competences, with innovative rules, on organisation, relation, and liability, which operate together with the traditional rules on production and products, and which express the *proactive innovative* approach of the European legislator.

As well observed by an authoritative scholar, Regulation No 178/2002 operates on different levels: 'that of law sources, ... [where it] introduces general principles ... establishes rules directly applicable, which do not need national execution measures; ... that of the institutional design to be adopted by any Member State; ... that of cooperation to provide among national organizations and Community organisation'[87].

This Regulation, in particular, expressly declares its systemic nature of act which 'lays down the general principles governing food and feed in general, and food and feed safety in particular, at Community and national level'[88], providing principles of general interpretation, operating as general uniform criteria applicable both to new rules and to the existing ones, at Community and national level, in the perspective of unified regulation.

Among the many relevant principles, definitions, and rules directly applicable, it is significant to mention the general definition of 'food business' extended to include 'any undertaking, whether for profit or not and whether public or private, carrying out any of the activities related to any stage of production, processing and distribution of food'[89], covering all stages of the food chain, including the primary stage, as proposed in the Commission White Paper of 2000.

The definition makes express reference to any of the activities performed in any phase of the food chain, and therefore refers to a model of phase business, a business which receives qualification and regime not by reason of performing a comprehensive and homogeneous set of activities, but simply as a consequence of performing a single activity, even non homogenous to the others of the phase, but in any case potentially relevant.

Reference to a phase business is not unknown to other legal experiences. In Italy it is sufficient to mention the new definition of agricultural business, as introduced in 2001[90]. The category is usually linked to those of chain and net, assigning legal relevance to models until recently considered only by economic analyses.

---

[87] S. Cassese, *Per un'autorità nazionale della sicurezza alimentare*, Milano, 2002, 16.
[88] Article 1(2) Regulation No 178/2002; italics added.
[89] Article 3(1)(2); italics added.
[90] Legislative Decree No 228 of 18 May 2001, modifying Article 2135 of the Italian Civil Code.

The result is the emergence of a food business, which implies legal regimes, administrative models, rules of organisation and of activity, passing through the traditional distinctions.

In other words, the object of regulation (food) and the objective of regulation (health safety) induced the European legislator of Regulation No 178/2002 to take note that in this sector a point regulation, referring to single split categories of subjects, could not be effective, and that it is necessary to adopt comprehensive rules, which qualify the actors not by reason of abstract categories, but simply by reason of their participation (whatever it is) in any production, processing or distribution activity.

Private food business actors receive rules, but at the same time they also operate as rule-makers, of self-regulation and self-responsibility, in a dialogue with public actors, which takes food safety as basic criterion of the legal statute of competition in the food market.

It follows that Regulation No 178/2002 has a twofold nature. Partly, it expresses an innovation-reaction (with the related conception of the ineffectiveness of an approach based exclusively on a response), inasmuch as it seems to delegate to technology a series of choices which, by their very nature, call for comparisons and hierarchies of interests that cannot always be solved only on the technical plane. But on the other hand, it also expresses an innovation-action, inasmuch as it introduces novel regulatory models, which lead to innovative institutional structures.

1.10 The reformed CAP

In the last two decades European discipline of agriculture has been characterized by a continuous process of reform, operating both on the economic strategy and on the legal framework.

At the opening of the '90 the Mac Sharry Reform marked a new approach, introducing a system of payment based on the dimension of land and not on the quantity of product. In the following years, Agenda 2000, the regulation on rural development of 1999, the MTR of 2003, and the new regulation on rural reform of 2005, in significant correspondence with the enlarging of the European Community with new member States, introduced a large (and still not concluded nor entirely defined ex ante) series of rules, both communitarian and national, which as a whole compose a European discipline of agriculture, which is common with reference to the goals, the institutions, and the tools adopted, but which does not imply a uniform '*mise en ouvre*' in the different Member States.

During this process, two phenomenon induced specific interest toward models and institutions of government of the agricultural sector:

- On one side, in most European countries we are facing a strong demand of local regulation, in the shape of self-regulation on a private negotiated and contractual basis, and in the shape of regulation coming from local and regional political entities.
- In the same time, European agriculture legislation is increasingly adopting a policy, which tries to find answers to the challenges coming from the enlargement with new Member States and from the evolution of negotiation on world trade, through a model which integrates a systemic dimension of regulation and a specific attention for the themes of governance, in a perspective which by itself implies the re-localisation of some decisions, which no more can be effectively taken only at central level.

This new approach is at work in different areas, including: economics, governance and distribution of powers, as well as relations among individuals. The result is a new model of '*plural*' regulation, a '*flexible droit*'[91].

European binding rules dealing with the environment, food safety, food quality, animal and plant health, animal welfare, traditionally operating as external limit to all the economic activities (external limit, which by its nature is assisted by criminal and administrative sanctions), starting from the reform of the CAP of the new century operate as internal habit of the agricultural activity, qualifying this activity and moving toward agricultural structures coherent with a profitability, which cannot abide from the satisfaction of environment, quality and safety goals.

Within this perspective, largely influenced by the evolution of international rules of trade, economic and productive activities are subject to legal rules, but at the same time are entitled to adopt voluntary and contractual rules, of self-regulation and self-responsibility, in a multilevel dialogue with multiple regulators, which assume as basic principles of action the value assigned to the autonomy of choices and decisions and the legal statute of competition in the market.

This European way to make law could be defined as a kind of government of procedural pragmatism, substituting the primacy of institutions to the primacy of sources of law[92]. It adopts definitions, introduces principles, chooses (and in certain cases creates) actors, public and private, national and communitarian, and then assigns to following acts and experiences, not only communitarian, but even national, regional and local, the construction of rules of daily life.

[91] See J. Carbonnier, *Flexible* droit, 7th ed., Paris, 1992.

[92] See M.R. Ferrarese, *Il diritto al presente. Globalizzazione e tempo delle istituzioni*, Il Mulino, Bologna, 2002.

The new Regulation No 1698 of 2005 on rural development[93] confirmed the path toward a European Agricultural Law, which is a kind of *ius commune*, where multilevel regulation is the tool to answer to variety and difference.

Market assumes a central regulatory role, and – as with reference to the path toward unique market for others productive sectors – even in agriculture the rules of competition involve critical application. Regulation No 1782/2003 not by accident underlines expressly the goals 'to ensure equal treatment between farmers and to avoid market and competition distortions'[94], as guidelines for implementation of general rules by national regulators, together with the principles of autonomy and freedom of operators.

The other coin of regulation, which integrates the central role assigned to market and autonomy, is cross-compliance. The freedom of operators with reference to model of structure and to content of activity (including the non-production), is balanced by the duty to conform to binding provisions regarding the environment, animal wealth and good agronomic conditions.

Cross-compliance becomes an internal way of life of agricultural holdings[95], and food quality is expressly located within the measures to increase competitiveness, confirming the intimate relation between agricultural law and food law. In the economic perspective, the shift is from a redistribution policy to a regulatory policy.

In the legal perspective, the system of national ceilings and the assignment to Member States of powers which allow different choices, require the adoption of common institutes, which involve even areas (like those of private law) until now left to national disciplines.

1.11 The regulation on accreditation and market surveillance

Another recent aspect of institutional innovation, that has become a focus of attention, in which agricultural law and food law confirm themselves as testing grounds for innovative legal institutes and for models endowed with expansionary capacity, has arisen in the framework of the adoption of the new rules on

[93] Council Regulation (EC) No 1698/2005 of 20 September 2005 on support for rural development by the European Agricultural Fund for Rural Development (EAFRD).
[94] Article 42 Regulation No 1782/2003.
[95] See the long list of Directives and Regulations included in Annex III of Regulation No 1782/2003, which refers to European rules existing from years but not sufficiently implemented through administrative or criminal sanctions, and which assigns to them a specific role of conformation of agricultural activity.

accreditation and market surveillance[96]. Regulation No 767/2008 of the European Parliament and of the Council, which was approved on 9 July 2008 and entered into force on 1 January 2010[97], regulates in a unitary manner a number of issues concerning the placing of products on the market: the accreditation systems of the certification bodies, the surveillance systems of the internal market and for control of products imported from third Countries, and the '*CE*' labelling system[98].

The original proposal drawn up by the Commission expressly excluded food products, as defined by Regulation No 178/2002, from its field of application, but during the European Parliamentary sessions the decision was made to limit the exclusion of food products to the part of the Regulation that concerned control over the internal market and over imported products. In contrast, the field of application was extended so that it would include the new measures on accreditation to all products, including food products.

The adoption of a unitary system of accreditation, on a public basis, of all the certification organisms, whatever the nature of the product to be certified, goes in the direction of a unitary approach to the market. Conversely, the exclusion of food products from the field of application of the new rules on surveillance of the markets and of imported products should not be seen as expressing a category-based segregation: rather, it is simply due to the previous existence of a specific (and by now well established) system of surveillance of the food market governed by Regulation No 178/2002. In other words, Regulation No 765/2008 extensively takes up again and extends to other product classes the models of surveillance and intervention in the market which some years ago were formulated and were introduced in the framework of food safety. Accordingly, in reference to the crucial theme of identifying the subjects, the competences and procedures entrusted with guaranteeing market protection for pre-eminent interests and values, it can be seen that food law, authoritatively recognised as a 'model of European unification'[99], increasingly proves to be a vehicle of institutional innovation.

---

[96] Regulation (EC) No 765/2008 of the European Parliament and of the Council of 9 July 2008 setting out the requirements for accreditation and market surveillance relating to the marketing of products and repealing Regulation (EEC) No 339/93. For an analysis reflecting the views of different scholars, under different perspectives, of the types of legal innovation resulting from this Regulation in terms of safety, control and responsibility, see M. Tallacchini, F. Albisinni, S. Amorosino, F. Capelli, F. Di Porto, N. Rangone, A. Moscarini, L. Ammannati, P. Borghi, L. Russo, Viterbo Congress, 2-3 December 2011, AIDA-Associazione Italiana di Diritto Alimentare, 'Controlli, certificazioni, responsabilità', in *Riv.dir. alim.*, www.rivistadirittoalimentare.it, n. 4-2011, 1-2012.

[97] Article 44 Regulation 767/2008.

[98] Article 1 Regulation 767/2008.

[99] L. Costato, *Il diritto alimentare: modello dell'unificazione europea*, in *Riv.dir.alim.*, www. rivistadirittoalimentare.it, n. 3-2009, 1.

1.12 The new scenario: European codes

It seems appropriate to conclude that the most recent European reforms, adopted before and after the entry in force of the Lisbon Treaty, move toward the progressive implementation of European Codes, of agriculture and food, which are common and shared codes, where needs and subjects, national, regional, and local, hold a relevant place, integrating themselves with decisions and options assumed at central level.

Those common codes of cand food law design a model of implementing European unity, which assigns greater value to the legal tools of regulation, compared to the traditional recourse to economic tools per se considered; a model of deeper integration between local and central levels, which is expanding also in other areas of the European experience, first of all that of the financial markets, where the present long lasting crisis induces to re-discover institutes and tools which have been originally elaborated as an answer to the crisis of food safety.

It is a model of codification much different from that of the national codes of 19th century, which were codes of separation. It is rather a process moving toward a *Corpus iuris*, which in some ways may recall the Justinian Code, reassembling the past while sketching the future, in a framework where the sources of law are different and plural. In this process of unification through law, an essential role is assigned to the legal interpretation, to the *ius dicere*, in all its forms and manifestations, judicial, administrative, academic, and practical.

Lawyers and lawyers associations (like EFLA, which meritoriously organised this international Congress) are called to a crucial contribution in this sensitive process, due to the same nature of the tasks entrusted to them.

2. Private law and agricultural food law

Preventing exclusion of small-scale-farmers through private standards in the global food chain – a research agenda[100]

Kai Purnhagen, Cerkia Barnard, Bernd van der Meulen and Harry Bremmers

2.1 Introduction

At its 18th international scientific congress held in Amsterdam, the European Food Law Association coined 'private food law' as a distinct area of food law in its own right.[101] As food law does in general, private food law covers the entire food chain from 'farm to fork'. At closer inspection, private standards are often set by actors at the end of the food chain to regulate behaviour at the beginning of the food chain. In private regulation, a distinctive part of governance, food law and agricultural law truly meet. It is self-made law where the primary sector often is the regulated party. In this contribution we propose a research agenda particularly addressing the situation where private standards are set by Western stakeholders (mainly retail) and used to regulate agriculture in the global South. The role of private governance in global food chains is specific because it is largely beyond the scope of state law. This contribution does not have the objective to provide any answers, but to find the right questions for future research.

In June 2013, the UN Food and Agriculture Organization (FAO) and the Organization for Economic Cooperation and Development (OECD) in their Agricultural Outlook 2013-2022 announced an increase of food prices from 10-40% in the coming years.[102] Despite the benefits that developing countries gathered by switching from understanding food markets as a market driven rather than a policy driven sector, production shortfalls, price volatility and trade disruption remain threats to global food security.[103] In perfectly operating markets prices are determined under conditions of complete transparency, homogeneity of input and output, existence of multiple suppliers (no power disequilibrium) as well as free market access and exit possibilities. Such 'ideal' conditions will never be met. Increasingly, agricultural market outcomes are influenced by actors downstream the supply chain, which control access to the final consumer. However, supply chain relationships are

[100] This chapter is a preliminary version of an article which will be published in the European Journal of Risk Regulation

[101] See the EFLA-book: B.M.J. Meulen (ed.) Private food law. Governing food chains through contract law, self-regulation, private standards, audits and certification schemes, Wageningen Academic Publishers 2011, available at: www.wageningenacademic.com/eifl-06.

[102] OECD-FAO Agricultural Outlook: 2013-2022. Press Release 06/06/2013, available at www.oecd.org/site/oecd-faoagriculturaloutlook/oecd-fao-expect-slower-global-agricultural-production-growth.htm.

[103] OECD-FAO Agricultural Outlook: 2013-2022. Press Release 06/06/2013, available at www.oecd.org/site/oecd-faoagriculturaloutlook/oecd-fao-expect-slower-global-agricultural-production-growth.htm.

only enduring in the long run if they guarantee the provision of income, foster innovation and stimulate economic activity. Prices serve as indicators for the value of food products. As to the OECD:

> High food prices are an incentive to increase production and we need to do our best to ensure that poor farmers benefit from them. Let's not forget that 70% of the world's food insecure population lives in rural areas of developing countries and that many of them are small-scale and subsistence farmers themselves.[104]

Real prices of raw material for food inputs show a constant decline since 1962[105]. The constant pressure on prices and the economic crisis have aggravated the problem of 'fair' pricing and income distribution along supply chains. Global sourcing of inputs in supply chains orchestrated by western companies creates a lack of transparency with respect to the income effects for farmers in developmental countries. Lack of transparency of origin of inputs as well as of conditions during their production complicates governance on multiple aspects: economic, sustained production and equity.

The OECD-FAO calls on governments that they 'need to find ways to give farmers access to technology to increase output and get more of their crops to market.'[106] Furthermore, '(g)overnments need to create the right enabling environment for growth and trade.'[107]

We agree with OECD-FAO regarding the need to secure especially small scale farmers (SSF) access to food markets. In terms of their policy recommendation to call on governments they seem to miss one important point. In a market-driven food sector the importance of governments as regulators decreases and is increasingly

[104] J.G. Da Silva, (2013). Food and Agricultural Organization of the United Nations: a statement by FAO Director- General. OECD-FAO Agricultural Outlook: 2013-2022. Press Conference.

[105] FAO, (2004) The State of Agricultural Commodity Markets 2004, annex 1, Commodity Prices in Real Terms. FAO: Rome.

[106] OECD-FAO Agricultural Outlook: 2013-2022. Press Release 06/06/2013, available at www.oecd.org/site/oecd-faoagriculturaloutlook/oecd-fao-expect-slower-global-agricultural-production-growth.htm.

[107] OECD-FAO Agricultural Outlook: 2013-2022. Press Release 06/06/2013, available at www.oecd.org/site/oecd-faoagriculturaloutlook/oecd-fao-expect-slower-global-agricultural-production-growth.htm.

replaced by retailers as private standard setters and enforcers.[108] This is the ethos of the rise of market states, which is characterised by decreases of regulatory potential of nation-state governments and rises of private regulators.[109] It is not effective to only rely on traditional government regulation based on classic accountability models of parliamentary democracy. Authorities in developed countries do not have a say over the effects of private standards that take place in developing countries, nor is it likely that SSFs will turn to courts or other institutions far from their place of residence. Governments in developing countries hold no power over standard setters located abroad. In order to successfully govern the global food sector, market regulatory models have to take the observations of market-state theory into account. In the food sector, regulation is hence only effective if they take into account firstly that retailers replace governments as regulators and secondly that regulation switches from classic top-down models to the provision of incentives. In line with the concern raised by the UN Special Rapporteur on the Right to Food on the role of retailers in realising the right to food,[110] research is needed that addresses the issue of accountability and legitimacy in the context of the rise of market-states with the aim to protect SSF against exclusion from global and local food markets. In order to cope with market-state challenges, emphasis needs to be placed on the corporate food regime with a focus on the development and enforcement of retailer standards in particular. Data is needed to verify the generally shared conviction that retailers and other food businesses supplement and sometimes even replace traditional governmental models of regulation (see Section 2). Highlighting the potentially adverse economic and human rights impacts of privatisation of regulation in global food systems, it is then important to develop an integrated market-state model assuring accountability and legitimacy of private standard setting. Accountability and legitimacy models in global private food chains have to be based on a combination of public and private governance instruments in market-state theory, particularly with a view

[108] See the example of the EU internal R. van Gestel, and H.-W. Micklitz, (2013). European integration through standardization: how judicial review is breaking down the club house of private standardization bodies, Common Market Law Review 50; and the observations by F. Cafaggi (2010). Private Regulation, Supply Chain and Contractual Networks: The Case of Food Safety. RSCAS 2010/10 (Robert Schuman Centre for Advanced Studies - Private Regulation Series); and S. Henson, and J. Humprey (2009). The Impacts of Private Food Safety Standards on the Food Chain and on Public Standard-Setting Processes: Paper Prepared for FAO/WHO. Codex Alimentarius Commission. Thirty-second Session. FAO Headquarters, Rome; and S. Henson and J. Humprey, (2011). Codex Alimentarius and private standards. Chapter 5 in: B.M.J. van der Meulen (ed.) Private Food Law. Governing food chains through contract law, self-regulation, private standards, audits and certification schemes, Wageningen: Academic Publishers, who argue that private standards have become the key driver and primary determinant of participation in global food supply chains.

[109] See D. Patterson and A. Afilalo (2008), The New Global Trading Order: The Evolving State and the Future of Trade, Cambridge Univ. Press.; B.M.J. van der Meulen (ed.)(2011) Private Food Law. Governing food chains through contract law, self-regulation, private standards, audits and certification schemes, Wageningen: Academic Publishers; K.P. Purnhagen (2013) The Politics of Systematisation EU Product Safety Regulation- Market, State, Integration, and Collectivity, Springer New York.

[110] UN Special Rapporteur on the Right to Food on the role of retailers in realising the right to food (2009).

on incentivising private regulators to adopt and enforce standards that ensure SSF participation in the global food market (see Section 3).

2.2 Privatisation: retailers as regulators through standards

The global agro-food regime has undergone significant transformation in recent decades. Two significant features of this transformation are the corporatisation of the agro-food system and the rise of retailers as powerful actors in global supply chains.[111] This is accompanied by increased privatisation of food regulation, in particular through the emergence of private food standards. Over the past three decades, private food standards have been so widely adopted by retailers in particular, that they have emerged as the primary mode of governance in global food supply chains.[112] While voluntary by contract law theory, the widespread adoption of private food standards means that these standards have become de facto mandatory as a precondition for participation in the majority of global supply chains.[113] While this initially applied primarily to export markets, retailer capture of food markets across the developing world in recent years has left few alternatives, even in a domestic context.[114] With private standards compliance increasingly being the primary determinant of market access in global food supply chains and with the development of these chains worldwide, these standards on the one hand carry the potential to ensure participation of SSFs in the global food chain as the conditions for participation is made transparent. However, they conversely also create a threat to SSF participation, if the respective SSF cannot meet the requirements of the standard. In this respect, a growing body of evidence highlights the risk that retail standards as the key determinant of market access, may exclude in particular SSF in the developing world from market access.[115] Most current private standards are geared towards large quantities of produce of uniform quality. Costs of audits and certification are generally borne by the producer. The investments involved are often unattainable for SSF. When excluded from global food supply chains, SSF who are dependent on agriculture for their household income increasingly risk losing their ability to earn a living.

[111] F. Snyder (2004). Economic globalisation and the law in the 21st century. In: A. Sarat (ed.). The Blackwell companion to law and society. Oxford: Blackwell publishing.

[112] L. Fulponi (2007). Private voluntary standards in the food system: the perspective of major food retailers in OECD countries. Food Policy, 31; S.J. Henson (2007). The role of public and private standards in regulating international food markets. Journal of international agricultural trade and development 4(1).

[113] S. Ponte and P. Gibbon (2005). Quality standards, conventions and the governance of global value chains. Economy and society, 34(1); B.M.J. van der Meulen (ed.)(2011) Private Food Law. Governing food chains through contract law, self-regulation, private standards, audits and certification schemes, Wageningen: Academic Publishers.

[114] T. Reardon, P. Timmer and J. Berdegue (2004). The rapid rise of supermarkets in developing countries: Induced organisational, institutional and technological change in agri-food systems. Journal of Agricultural and Development Economics, 1(2): 15.

[115] J. Berdegué, F. Balsevich, L. Flores, T. Reardon (2005), Central American supermarkets' private standards of quality and safety in procurement of fresh fruits and vegetables. Food Policy 30:245-269.

Private food standards have thus emerged as important threat to household income and food security and in this way potentially constrain the realisation of the right to food and other human rights such as the right to development and non-discrimination.[116]

The potential human rights concerns of standard setting as a private regulatory activity necessitates research on how to ensure greater accountability for private regulators that fall outside of the constitutional accountability frameworks within which public regulators operate.

2.3 Providing incentives to effectively ensure SFF access to global food markets

In the developed world, a fragmented regulatory space has emerged with governance procedures such as private standard setting supplementing more traditional command and control measures by governments.[117] In turn, this allows applying traditional government regulation such as competition law to adverse effects of private governance.[118] Effective regulation aiming at closing gaps in governance may be tackled with classical governmental means such as top-down regulation.[119] However, even in developed countries, there is an increasing trend for substitution of classical command-and-control through incentive mechanisms.[120] In developing countries this trend is even more evident as governments often lack in effectiveness. As traditional governmental regulation such as liability and competition law is either ineffective, imprecise or simply not available, governance procedures such as private standard setting often form the last resort for effective regulation. This increases the importance of private standards in these areas significantly, which have the highest potential to create some form of rule of law.[121]

The movement of global administrative law (GAL) has been occupied with researching these phenomena from a perspective of accountability. In line with our claim, GAL proposes to switch the perspective from government to corporate

[116] UN Special Rapporteur on the Right to Food on the role of retailers in realising the right to food (2009).
[117] C. Möllers, (2006). European Governance – Meaning and Value of a Concept, Common Market Law Review 43.
[118] See H. Schepel (2005) The Constitution of Private Governance- Product Standards in the Regulation of Integrating Markets, Oxford: Hart Publishing; F. Stancke (2011). The limit of private food law – Competition law in the food sector. Chapter 17 in: B.M.J. van der Meulen (ed.) Private Food Law. Governing food chains through contract law, self-regulation, private standards, audits and certification schemes, Wageningen: Academic Publishers.
[119] C. Möllers, (2006). European Governance – Meaning and Value of a Concept, Common Market Law Review 43.
[120] K.P. Purnhagen (2013) The Politics of Systematisation EU Product Safety Regulation- Market, State, Integration, and Collectivity, Springer New York.
[121] See on the difference between governmental de-jure standards and private de-facto standards J. Farrel and G. Saloner (1986). Installed Base and Compatibility: Innovation, Product Preannouncements, and Predation, American Economic Review, 76:940-955.

standard setting as corporations exercise 'de facto regulatory functions'[122]. According to GAL, corporations shall incorporate and enforce a set of principles and values identified through a 'shared set of norms or norm-guided practices'[123] to ensure accountability of emerging global governance mechanisms.[124] As such, GAL has been defined as including:[125]

> the legal mechanisms, principles and practices, along with supportive social understanding, that promote or otherwise affect the accountability of global administrative bodies, in particular, by ensuring these bodies meet adequate standards of transparency, consultation, participation, rationality and legality and by providing effective review of the rules and decisions these bodies make.

In line with Curtin and Senden,[126] accountability comprises two features:

1. a 'normative concept, a set of standards for the evaluation of the behaviour of public actors'. In this understanding, accountability is focussed on 'normative issues, on the assessment of the actual and active behaviour of [...] agents' and
2. a '[...] mechanism, as an institutional relation or arrangement in which an actor can be held to account by a forum'. This understanding of accountability focusses on the institutions which can hold entities accountable, instead of focussing on whether entities have acted in an accountable manner.

While we share the view of a need to switch perspective to an understanding where corporate bodies as standardisation companies work as de-facto regulators, we find it doubtable whether these views on accountability, when adopted without modification, can really help to improve market access for SSF's at both the level of normative standard setting and of control. The weaknesses of the models proposed by GAL and Curtin and Senden are vested in the fact that they still embody a touch of 'stateness', which cannot be brought in line with market-state challenges that arise in the increasingly privatised food sector.

[122] D. Richemond-Barack, (2011). Regulating war: A taxonomy in global administrative law. The European Journal of International law, 22(4); L. Busch, (2011). Quasi-states? The unexpected rise of private food law, Chapter 2 in: B.M.J. van der Meulen (ed.) Private Food Law. Governing food chains through contract law, self-regulation, private standards, audits and certification schemes, Wageningen: Academic Publishers.

[123] D. Richemond-Barack, (2011). Regulating war: A taxonomy in global administrative law. The European Journal of International law, 22(4).

[124] N. Krisch, (2006). Global Administrative Law and the Constitutional Ambition. LSE Law, Society and Economy Working Papers 10/2009. London School of Economics and Political Science Law Department.

[125] B. Kingsbury, N. Krisch, R.B. Stewart and J.B. Wiener (2005). Foreword: global governance as administration – national and transnational approaches to global administrative law. Law and Contemporary Problems, 68: 5.

[126] D. Curtin and L. Senden (2011). Public Accountability of Transnational Private Regulation: Chimera or Reality? Journal of Law and Society, 38(1).

In view of the limitations to the explanatory value of current laws and theory including GAL and their ability in practice to deal with the adverse impacts of standards on the human right to food of SSF, research needs new pathways. We argue that, as private standard setters are emerging as de-facto regulators, research requires a fresh view on issues of legitimacy, accountability and enforcement. The privatisation of the food regime requires taking into account insights on legitimacy and accountability from private law as well, which fill the gaps of effective government enforcement. Private law mechanisms hence supplement and complement traditional public law models of accountability, legitimacy and enforcement. Our hypothesis, therefore, is that only the combination of both, public and private mechanisms in an integrated model can work towards the end of effective oversight of standards in the global food regime and of practices within the framework of these standards.

In this respect research is needed which explicitly maps and addresses governmental possibilities and failures in standard oversight. This includes governments both in countries where the producers are located that are affected by private standards, and in countries where the retailers are located that export obligations through the private standards they set. By consequence territorial as well as extra territorial obligation of governments are to be considered.[127] Particular emphasis should be placed on the emergence of standard setters as de-facto regulators and their impact on the realisation of the right to food. While there is a strong presumption that private standards, which primarily aim at ensuring food safety, exclude SSFs from the global food market, hands-on data is still missing. What is therefore strongly needed is the collection of data about the impact of food standards of private organisations such as GlobalGAP, SQF, UTZ or Fairtrade on SSFs.

Furthermore, in order to understand how we may make private law mechanisms work for filling the gaps so identified, research is needed how bargaining processes impact the standard setting and their oversight. From a private law perspective, standard setting and their enforcement has to be viewed in light of contract governance,[128] where parties and stakeholders in a (quasi) contractual network continue to bargain about standards from the point of standard setting until enforcement. Applications of insights from contractual governance is needed that highlights how models of private law may be used to supplement or substitute gaps resulting from governmental failure in the control of standard implementation.[129]

[127] U. Hausmann, (2009) The hungry challenging the global elite, in: O. Hospes and B.M.J. van der Meulen (eds.). Fed up with the Right to food? The Netherlands' Policies and Practices Regarding the Human Right to Adequate Food. Wageningen: Wageningen Academic Publishers.

[128] F. Moeslein, Contract Governance within Corporate Governance – A Lesson from the Global Financial Crisis, in: F. Cafaggi, S. Grundmann (eds.), From Exchange to Cooperation in European Contract Law, 2013.

[129] see for a first attempt S.M. Maurer, Public Problems, Private Answers: Reforming Industry Self Governance Law for the 21st Century, on file with author.

Particular emphasis must be placed on understanding standard setting and enforcement behind the theory of relational contracts and their external effects, which has to date been researched mainly in the area of company law. As standards are only negotiated against the social background of the standard setters,[130] insights of the social web of contracts which govern companies may also help to understand the social web of standards. It is hence important to understand what incentivises the respective contracting partners in order to understand how private standards seen as relational contracts are drawn and enforced. This has to be seen particularly with regards to negative externalities such as exclusion of SSFs. In this respect, such relational contracts are legitimate and effective if partners are able to secure mutual trust in each other by securing compliance through incentives. In view of contractual governance, one may even express the opinion that such relational contracts are only legitimate if the external effects created from the contractual relationship do not exceed a certain measure.

In order to successfully make use of regulation via private standard setting, particularly in the food sector, research is needed to understand the network behaviour of the participants.[131] If we gain more insights on how the bargaining process actually works, we may also know how we may incentivise the parties in order to realise the human right to food. Such incentivising may supplement or even substitute the missing enforcement elements in private governance of accountability and legitimacy which are rooted in traditional state-based models.

2.4 Conclusion

Compliance with private standards is rapidly becoming the decisive factor distinguishing inclusion and exclusion from the market for agricultural producers. If Africa would manage to increase its share of global exports by just 1%, it would generate about $ 150 Billion annually. This is four times what it receives in development aid.[132] Agriculture holds the largest potential to come closer to realising this ambition. For SFFs to receive a fair share or even to just survive in the globalising world, they need to have access to the local and global food markets. Research is needed that helps to counter the exclusionary effects of the private standards that rule these markets. In absence of effective (state) governmental control, private standards and their governance shall be turned into tools to contribute to this

[130] See R. Axelrod, W. Mitchell, R. Thomas, D.S. Bennett, E. Bruderer, Coalition Formation in Standard-setting Alliances, 1995. Management Science, 41:1493-1508.

[131] F. Cafaggi (2010). Private Regulation, Supply Chain and Contractual Networks: The Case of Food Safety. RSCAS 2010/10 (Robert Schuman Centre for Advanced Studies – Private Regulation Series).

[132] Message delivered in the popular science movie 'Beauty and the Beef. Achieving compatibility between wildlife conservation and livestock development in southern Africa; produced by M&M Pictures for the Department of Veterinary Tropical Diseases, Faculty of Veterinary Science, University of Pretoria.

objective. In this sense, research is needed that investigates the respective incentive structures in bargaining over standards and their enforcement.

We call upon or friends and colleagues in the European Food Law Association and beyond to join us in this endeavour.

3. The new Common Agricultural Policy – objectives relevant to food law

Donald Rennie

Let us go back to first principles. The Common Agricultural Policy is as old as the European Union, or perhaps I should use its original title, the European Common Market. It was the first common policy to be put into effect.

Its origins are to be found in Article 39 of the Treaty of Rome of 1957. This states that the objectives of the common agricultural policy shall be:

- to increase agricultural productivity by promoting technical progress and by ensuring the rational development of agricultural production and the optimum utilisation of the factors of production, in particular labour;
- thus to ensure a fair standard of living for the agricultural community, in particular by increasing the individual earnings of persons engaged in agriculture;
- to stabilise markets;
- to assure the availability of supplies;
- to ensure that supplies reach consumers at reasonable prices.

These objectives were set in the 1950s against a background of food shortages across Europe following the terrible destruction and disruption of the Second World War. Conditions in Europe improved steadily from those lean years and in the period of unexampled prosperity before the economic crash of 2008 the policy came to be perceived as wasteful. Now we are facing both poverty and economic malaise across Europe and a world whose population is expected to increase from 7 billion to 9 billion in the course of the next few years. The issue which caused so much anxiety to the framers of the Treaty of Rome and which is now called 'food security' has come to the fore again.

We must therefore examine the details of the treaty obligations incumbent on the Member States of the European Union and consider the extent to which the proposed changes to the CAP are compatible with those obligations and the impact which they will have on Food Law.

We also have to consider that the Policy is now to be implemented in view of Article 11 of the International Covenant on Economic, Social and Cultural Rights, which entered into force on 3 January 1976 and which provided:

1. The States Parties to the present Covenant recognise the right of everyone to an adequate standard of living for himself and his family, including adequate food, clothing and housing, and to the continuous improvement of living conditions. The States Parties will take appropriate steps to ensure the

realisation of this right, recognising to this effect the essential importance of international co-operation based on free consent.

2. The States Parties to the present Covenant, recognising the fundamental right of everyone to be free from hunger, shall take, individually and through international co-operation, the measures, including specific programmes, which are needed:
   a. To improve methods of production, conservation and distribution of food by making full use of technical and scientific knowledge, by disseminating knowledge of the principles of nutrition and by developing or reforming agrarian systems in such a way as to achieve the most efficient development and utilization of natural resources;
   b. Taking into account the problems of both food-importing and food-exporting countries, to ensure an equitable distribution of world food supplies in relation to need.

Professor Bernd van der Meulen, in his contribution to 'Roadmap to EU Food Law',[133] correctly describes this provision as 'the most fundamental dimension of international food law'.

When the programme of the Congress was in course of preparation it was anticipated that the shape of the new Common Agricultural Policy would be clear by now but this was not the case. The amount of money available to implement the policy would not be known until the European Union had decided its final budget. On 30 August 2012 the Cyprus Presidency convened an informal Council meeting. Bilateral consultations were held in advance of the meeting and as a result of these the Cypriot Presidency presented an informal summary which stated:

> Concerning the Common Agricultural Policy, given the sharply opposite preferences of the Member States, '*the contribution to the overall reduction has to come from both Pillars [that is to say, both Pillar 1 and Pillar 2 which I discuss later] and further discussion is needed regarding the relative priority between them*'. At the same time, while respecting the distinct objectives of the two Pillars, it would seem appropriate to provide for somewhat more flexibility between them, possibly in both directions. The principle of a more equitable distribution of direct support is accepted, while further consideration is required on the ambition and speed of convergence as well as on degressivity. The Presidency understands that the majority of Member States consider that the distribution key for rural development is a necessary part of the overall agreement on the MFF [i.e. the Budget].

[133] I. Scholten-Verheijen, T. Appelhof, and B.M.J. van der Meulen, Roadmap to European Food Law, Eleven International Publishing, The Hague 2011.

The shape of the reformed Common Agricultural Policy was eventually agreed in the dying days of the Irish Presidency on 26 June 2013. As is often the case, the agreement is a political fudge, with many very important details still to be worked out both at European and at national level.

The review is the first since the principle of co-decision was extended by the Treaty of Lisbon. Co-decision is now described as the EU's standard decision making process. The EU website[134] sets out clearly and concisely the procedure which is followed:

- The European Parliament and the Council review proposals by the Commission and propose amendments. If the Council and the Parliament cannot agree upon amendments, a second reading takes place.
- In the second reading, the Parliament and Council can again propose amendments. Parliament has the power to block the proposed legislation if it cannot agree with the Council.
- If the two institutions agree on amendments, the proposed legislation can be adopted. If they cannot agree, a conciliation committee tries to find a solution. Both the Council and the Parliament can block the legislative proposal at this final reading.

While this is intended to address to some extent the undoubted democratic deficit of the European Union it does lead to a more complex decision making process and in consequence to extra delay. As at 30 August 2012 the members of the European Parliament had submitted some 7,500 amendments to the proposed regulations and these will have to be voted upon in order the determine the Parliament's position so that the co-decision procedure may get underway.

We must examine the proposed new policy to ascertain to what extent it is compatible with the original treaty structure.

The primary obligation under the Treaty is to formulate a policy which will 'increase agricultural productivity'. All the other objectives of the policy are subordinate to this and are intended as examples of the means by which the Policy is to be achieved.

The treaty sets out three means to this end:

- The first is by promoting technical progress. The range of technical progress since the 1950s has been immense, involving improvements to livestock and crop production, but also the use of advanced machinery and buildings more proof against vermin. I suggest that technical progress should also include the investigation and use of genetically modified organisms where found to be safe and beneficial.

---

[134] See: http://europa.eu/about-eu/basic-information/decision-making/procedures/index_en.htm.

- The second is by ensuring the rational development of agricultural production. This is a very subjective criterion, involving value judgments on the part of the legislator as to what is 'rational development'. Does it mean, for example, that the maximisation of productivity must be carried out whatever the cost to the environment by way of loss of biodiversity and environmental degradation?
- The third is the optimum utilisation of the factors of production, in particular labour. This provision reflects the fact that the policy is a social as well as an economic one. The labour market now is very different from that which existed in 1957.

The intended consequences of implementation of the policy are then set out in Article 39, Paragraphs (b) to (e):

- The first intended consequence is 'to ensure a fair standard of living for the agricultural community, in particular by increasing the individual earnings of persons engaged in agriculture'. The background to this is that in the 1957 workers in agriculture were earning less than workers in heavy industry and there was therefore a fear that workers would leave the land to seek work in the cities. Formulating an agricultural policy which would keep workers on the land would be beneficial to the life of the countryside and would at the same time ensure that there was an adequate workforce to keep up the productivity of agriculture. The drift from country to town has been a fact of economic life since the Middle Ages and is a worldwide phenomenon.
- The second intended consequence is 'to stabilise markets'. Agriculture has always been at the mercy of weather, and shortage or oversupply can lead to wild fluctuations in product prices. The Policy was intended to remove some of this risk element from agricultural production. In practice various means were attempted to achieve this, such as intervention buying to maintain a floor price and the payment of direct subsidies to farmers to allow them to sell food at what would otherwise be a loss.
- The third intended consequence was to assure the availability of supplies. Given the background of a fear of starvation this was the most obvious outcome which the Policy sought to achieve.
- The fourth and last intended consequence was 'to ensure that supplies reached the consumer at reasonable prices'. The Arab spring was the result mainly of anger at the rapid rise in the price of basic foodstuffs and when that is considered it is obvious why the makers of the Treaty in 1957 stated that food should be available at reasonable prices.

These objectives and the intended consequences are not necessarily compatible with each other. In particular, cheap food for consumers and a fair standard of living for rural workers may not both be achievable without some measure of state subsidy or other intervention. This can be demonstrated by the discontent recently shown by dairy farmers in many parts of Europe who have complained that the power of milk processors and supermarkets means that the price paid to farmers is less than the cost of production of milk.

Before examining the detail of the proposed new CAP, I look at the definition of 'food' in the General Food Law (Regulation 178/2002). In terms of Article 2 of the Regulation 'food' (or 'foodstuff') means any substance or product, whether processed, partially processed or unprocessed, intended to be, or reasonably expected to be ingested by humans. But the Regulation goes on to state '"Food" shall not include: (1) feed; (2) live animals unless they are prepared for placing on the market for human consumption; (3) plants prior to harvesting' and various other items with which we are not here concerned.

Paradoxically, however, Article 3 provides that '"food law" means the laws, regulations and administrative provisions governing food in general, and food safety in particular, whether at Community or national level; it covers any stage of production, processing and distribution of food, and also of feed produced for, or fed to, food-producing animals'.

Equally paradoxically '"food business" means any undertaking, whether for profit or not and whether public or private, carrying out any of the activities related to any stage of production, processing and distribution of food.' One would have thought that even though the husbandry of live animals or the cultivation of crops prior to harvest do not fall within the definition of 'food' they are in fact stages in food production.

Following on from the definition of food business the Article defines 'food business operator' as meaning 'the natural or legal persons responsible for ensuring that the requirements of food law are met within the food business under their control'.

Although the majority of the activities of farmers are concerned with live animals and crops prior to harvesting, after the animals have been slaughtered and the crops harvested, farmers are subject to the same legal regime as others involved in the food chain. In addition, however, the Common Agricultural Policy requires that in order to qualify for subsidy the farmer must keep the land in Good Agricultural and Environmental Condition and must adhere to certain Statutory Management Requirements, relating to the environment and to public, animal and plant health and to animal welfare. These requirements and standards are collectively known as Cross Compliance.

As far as Food Law is concerned, the farmer is obliged to comply with Articles 14, 15, 17(1), 18, 19 and 20 of the General Food law, Regulation (EC) No 178/2002. As is to be expected these are the articles which deal with the marketing or attempted marketing of unsafe food and of the marketing and feeding of unsafe feed.

Article 14 provides that food shall not be placed on the market if it is unsafe. The article then goes on to define what is meant by unsafe food and further to define what is meant by food injurious to health and unfit for human consumption. There are additional provisions in the article expanding and explaining these simple principles.

Article 15 provides that feed, (that is material fed to animals) shall not be placed on the market or fed to any food-producing animal if it is unsafe. Feed shall be deemed to be unsafe for its intended use if it is considered to have an adverse effect on human or animal health or if it would make the food derived from food-producing animals unsafe for human consumption. Again there are additional provisions explaining and expanding upon these principles.

Article 17(1) provides that 'Food and feed business operators at all stages of production, processing and distribution within the businesses under their control shall ensure that foods or feeds satisfy the requirements of food law which are relevant to their activities and shall verify that such requirements are met.'

Article 18 deals with the traceability of food, feed, food-producing animals, and any other substance intended to be, or expected to be, incorporated into a food or feed. Traceability is to be established at all stages of production, processing and distribution so that food and feed business operators shall be able to identify any person from whom they have been supplied with a food, a feed, a food-producing animal, or any substance intended to be, or expected to be, incorporated into a food or feed. Food business operators are directed to have in place systems and procedures which allow for this information to be made available to the competent authorities on demand. Similarly, they are required to have procedures to enable them to identify businesses to whom the product has been supplied. The Article also makes provision for labelling to facilitate traceability, but plainly this is not applicable in the cases of live animals or of bulk crops and therefore alternative documentation must be in place.

Article 19 lays on food business operators various responsibilities to take steps to remove unsafe food from the market or, in the event that it has already been distributed to consumers, to recall it. Various positive duties are laid on food business operators to inform and co-operate with the food safety authorities with a view to ensuring that harm does not come to consumers of food.

Article 20 makes similar provision regarding feed business operators.

The consequence of the incorporation of these articles of the General Food Law into the Statutory Management Requirements of the Cross Compliance requirements of the Common Agricultural Policy is that the farmer will normally be a food business operator, even though the majority of his activities relate to animals and crops which fall outside the definition of 'food'. There may be some few farmers who do not produce food, for example if they produce only flax for the production of linen or biofuel or tobacco, but the main business of farms is the production of food for human consumption.

The normal arable farmer or stockman will therefore be in exactly the same position as businesses further down the supply chain so far as penalties for breach of obligations under and derived from the General Food Law and from domestic food safety legislation are concerned. But the farmer is at risk of additional and even more stringent penalties. The effect of cross compliance is that a farmer who fails in his obligations under Food Law is also penalised by loss in whole or in part of the subsidy which he would otherwise receive. The Commission proposal states that these penalties should be proportionate but farmers who may be penalised will think otherwise.

Mr George Lyon, a Member of the European Parliament and the shadow rapporteur for Direct Payments and CAP Financial Regulations, has called for proportionality of penalty. He has also proposed that there should be an early warning system for farmers who break cross compliance. He proposes that a communication should be sent to a farmer in breach giving him an opportunity of remedying the breach. This assumes, of course, that the breach is capable of remedy and is not the result of a deliberately fraudulent or otherwise criminal act. He also proposes that the farmer who has a good record of compliance with the regulations should be rewarded by a reduction in the number of inspections which he has to undergo, which would have the result of reducing the burden of bureaucracy which the has to bear. These proposals relate to all the requirements of cross compliance and not just to those which are relevant to Food Law.

The majority of the reforms which have been agreed have no impact on Food Law, for example, the changes to the rules on direct payments, the 'greening' proposals, the provision that is made for young farmers and the attempt to reach a satisfactory definition of an 'active farmer'. The last is intended to ensure that someone who has retired or withdrawn from farming does not continue to receive subsidy payments. The 'greening' of the policy is a means to sell the reformed Policy to a cynical urban electorate, although naturally this is not admitted.

The area of most concern to food lawyers is in what is described as 'market management mechanisms'. In their press release[135] the Commission state:

> Amendments to the Single Common Market Organisation (CMO) rules aim to improve the market orientation of EU agriculture in light of increased competition on world markets, while providing an effective safety net for

[135] See: http://europa.eu/rapid/press-release_MEMO-13-621_en.htm.

> farmers in the context of external uncertainties (together with direct payments and options for risk management under rural development). The existing systems of public intervention and private storage aid are revised to be more responsive and more efficient, for example with technical adjustments for beef and dairy. For dairy, these changes, come in addition to the 2012 'Milk Package' which is incorporated into the Regulation and strengthening the bargaining power of farmers.

In addition, the agreement provides:

> In order to improve farmers' negotiating position in the food chain, the Commission is looking for a better organisation of the sectors with a few limited derogations to EU competition law. Rules related to the recognition of Producer Organisations (POs) and inter-branch organisations are now covering all sectors – with further options for establishing such organisations now transferred to Rural Development funding. Furthermore, the possibility for farmers to collectively negotiate contracts for the supply of olive oil, beef, cereals and certain other arable crops is foreseen under certain conditions and safeguards. The Commission will provide guidelines about potential issues relating to competition law.

This provision is intended to address an anomaly in the food chain. The majority of consumers buy their food from a very small number of supermarkets. If farmers combine to seek to obtain countervailing economic power their activities are struck at by national and E.U. competition law. This proposal seeks to correct this imbalance. The details of how this proposal is intended to work are still to be provided.

The formal adoption of the reform by the European Parliament and the Council will come later in 2013 with a view to having the CAP reform in place as from 1 January 2014, although certain parts will not come into operation until much later.

Two points in particular should be noted:

- Firstly, it is difficult to reconcile the provisions for greening especially with Treaty obligations 'to increase agricultural productivity' and 'to ensure a fair standard of living for the agricultural community, in particular by increasing the individual earnings of persons engaged in agriculture'.
- Secondly, many issues are now left to national and regional authorities to determine. In that situation can one properly talk about a 'Common' Agricultural Policy?

4. The regulation of quality: beginning of food law[136]

Mariano López Benítez

The project 'Agenda 2000'[137] set as one of its fundamental goals the quality of products. Since then, quality has played a major role in the transformation of the legal system. Nevertheless, quality was by no means an unknown value neither in European Law nor in national legal systems, where some partial regulations on the topic could be found within food law. Still, the emphasis put on this topic since last century shows how important quality has become for both European and national legal systems.

In point of fact, quality assumes at present a significant role in the context of both a global economy and global markets: the main characteristic of European food products is quality. Notwithstanding, our aim here is to underline how quality affects in two different ways the traditional notion in our legal system and, specially, in our food law. First, we find what could be described as 'external influence', which has an impact on functions and powers. Secondly, it has influenced the actual organisation within the Administration, forcing to create a new infrastructure, whose aim is solely to be in charge of quality control. However, these phenomena are not exclusive to Food law, as the goal of quality has made itself present in other sectors within the legal system, where transformations due to quality are also taking place. In addition to this, it must be taken into account how quality assumes a horizontal function, thus reciprocally influences other sectors within the legal system to such an extent, that it can't be argued that all of them share the same infrastructure for quality (at the very least for its superior grades).

Focusing on food law, it must be considered that the agricultural part in 'Agenda 2000' set as one of its goals 'safety and food quality'. Those aspects were part of the administrate activity of limitation related to the traditional agricultural and food laws and implied regulation, control and surveillance powers, as the ones carried on to ensure quality for food products.

Nonetheless, as the new values of food safety and quality have expanded and reinforced, the power of the Administration has shrunken. That being so, the power to set quality standards for food products is quite restricted, as this is now

[136] Note from the editor: The regulation of quality has always been at the heart of Food Law and is one of its driving forces. In this chapter the author analyses briefly the importance of the regulation of quality in food law under the influence of Agenda 2000 and how this is affecting the structure of the Law Itself.
[137] COM (97) 2000.

a Communitarian competence, usually exercised by the implementation of CMO or specific regulations. There is very little left for the National Administrations.

It is however true that the Member States keep the competence of controlling that products comply with the requirements of Community legislation. Even so, this administrative control is residual, as producers themselves, on the one hand, and accredited agencies or associations (quite frequently private) on the other are increasingly been attributed this function by the EU law. This is clearly visible in both PDO and PGI products, whose distinctive signs describe the highest level of quality.

In the Spanish law system, since *Estatuto del Vino*[138] in 1932, quality control for PDO products has been carried out by *Consejos reguladores*[139]. The juridical form of those organisms has undergone changes throughout time, but always remaining as public entities. Not only did they verify the quality of products (specially that their elaboration met the set standards from those PDO products), but they also conducted inspection and sanctioning functions. Further, they were in charge of conferring guarantee certificates that ensured quality and identity of the product.

Yet Article 10 Council Regulation (EEC), no. 2081/1992, on the protection of geographical indications and designations of origin for agricultural products and foodstuffs, stated: 'An inspection structure may comprise one or more designated inspection authorities and/or private bodies approved for that purpose by the Member State'. This has been extended to other food sectors (like wine), until then exempted from quality requirements set for other food products. For all that, it does not seem that this legislation prevents public organisms from carrying out control and certification. They must, however, meet some additional requirements, which can be found in Norm EN-45011. Among them:

- A structure where members with executive tasks are elected among the representatives of interests within the process. It is enough that the structure can guarantee impartiality and allows every part to participate in the process.
- A permanent staff, in no case under control of those with a commercial interest in the certification in progress.
- A description of the process through which the organism obtains its certification.

This considered, there are several options under consideration. As some argue that Consejos reguladores and public organisms in general are now unable to keep exercising their former functions, some others believe that this is still possible, as long as they adjust their structure and working schemes to the new legislation (specially, there must be a clear separation between the organs in charge of control and those entrusted with regular administration). As regards this, the

[138] Wine Act.

[139] Regulatory Boards.

Spanish legislation is quite divergent. The reason for this is that the *Comunidades Autónomas*, which hold the regulation competences of PDO and PGI, are following dissimilar paths in respect to this.

Some regions have privatized the control: thus, a private entity is the one entitled to verify if the producer is respecting legislation. The entity elaborates a binding report with this information and hand it to the Administration or Consejo Regulador (considering that they are in charge of giving the quality certificate), or even the entrepreneur (who, in some cases and under the so called 'auto-control' can be in charge of giving the certificate).

On the other hand, some other regions have made minor changes. Therefore, *Consejos Reguladores* still play a major role, but now in the form of *Corporaciones de Derecho Público*[140], in which control and administration functions have been clearly separated. Additionally, *Consejo Regulador* must be accredited by ENAC[141], as stated in Norms EN 45004 and 45011.

Still, the main problem regarding the transformation in the quality control and guarantee of food products lays in the juridical nature of both the control and certification functions. Traditionally, the correspondence between organs and their functions was perfect: a public power was exercised only by a public organ or entity. However, this is no longer true, as private subjects are exercising public functions. A new difficulty then rises: how to ensure minimum legal safeguards in relation to functions exercised by private entities.

It must be taken into account that reports written by these private entities are imposed to the Administration. Consequently, I believe measures should be taken to ensure that every legal safeguard necessary to take any administrative decision (by the Public Administration) is also required of these private entities. The same must be said of the process by which a private control body gains accreditation from the National Entity.

In addition to this, the process by which a private control body gains accreditation from the National Entity needs to respect those legal safeguards. At the present moment, a lack of transparency and publicity, as well as serious weaknesses in the procedure, are being criticized by doctrine.

140 Public Law Bodies.

141 *Entidad Nacional de Acreditación,* Spanish National Accretidation Body.

5. Judicial protection of quality in the case of Parmigiano Reggiano

Giorgio Bocedi

5.1 Parmigiano Reggiano: a noble and historic designation of origin

Parmigiano Reggiano is a cheese made from cow's raw milk with a long history strictly and necessarily linked with its area of origin and it is famous today throughout the world so as to be considered 'The King of Cheeses' on the part of the specialist literature[142] and of most of consumers.

It is only made in a defined geographical area of Italy: indeed, both production of milk and its transformation into cheese take place in the historical provinces of Parma and Reggio Emilia (which give it the geographical name Parmigiano Reggiano), Modena, Bologna, on the left bank of the River Reno and Mantua, on the right bank of the River Po, the area of origin of the cheese Parmigiano Reggiano.

As regards its roots, Parmigiano Reggiano was first made (not just manufactured) several hundred years ago, more precisely eight centuries ago.

This cheese today is still identical to how it was eight centuries ago, made in the same places with the same expert ritual gestures, the same traditional process, through the same natural ingredients (cow's raw milk, calf rennet and salt), and having the same appearance and the same extraordinary fragrance.

The Parmigiano Reggiano has, *inter alia*, the following characteristics:

- the prohibition to use any kind of silage in the feed of the cows;
- the prohibitions to use any additive in the production step;
- a long and natural maturation, specifically a minimum maturing period of 12 months;
- a high content of proteins, vitamins and minerals that make it highly digestible.

5.2 The protection of the designation of origin Parmigiano Reggiano

The protection of the name Parmigiano Reggiano as designation of origin has been recognised over the years in the national and European union (EU) territories, and internationally, through different legal tools.

---

[142] Just an example: WIPO Magazine, 1/2011 www.wipo.int/wipo_magazine/en/2011/01/article_0005.html.

5.2.1 Italy

As regards the national protection through the following legislation:

- D.P.R. 18 November 1953, n. 1099 – Esecuzione della Convenzione di Stresa.
- Legge 10 April 1954, n. 125 – Tutela delle denominazioni di origine e tipiche dei formaggi.
- D.P.R. 30 October 1955, n. 1269 – Riconoscimento delle denominazioni circa i metodi di lavorazione, caratteristiche merceologiche e zone di produzione dei formaggi.
- D.P.C.M. 4 November 1991 – Estensione della denominazione di origine del formaggio 'Parmigiano Reggiano' alla tipologia 'grattugiato'.

5.2.2 European Union

When the European Union (at that time European Economic Community) legislator, within the framework of the CAP (Common Agricultural Policy), in 1992, introduced a *sui generis* system of protection of Geographical Indications (GIs) for agricultural products and foodstuffs[143], the name Parmigiano Reggiano was registered as Protected Designation of Origin (PDO) by Commission Regulation (EC) No 1107/96[144], which entered into force on 21 June 1996.

In 2006 Regulation (EC) 510/06 entered into force, repealing and replacing Regulation (ECC) 2081/92.

Presently, the name Parmigiano Reggiano is registered as PDO under Regulation (EU) No 1151/2012, of 21 November 2012, on quality schemes for agricultural products and foodstuffs, which entered into force on 3 January 2013[145], repealing and replacing Council Regulation (EC) No 510/06[146].

[143] Council Regulation (EEC) No 2081/92 of 14 July 1992 on the protection of geographical indications and designations of origin for agricultural products and foodstuffs.
[144] Commission Regulation (EC) No 1107/96 of 12 June 1996 on the registration of geographical indications and designations of origin under the procedure laid down in Article 17 of Council Regulation (EEC) No 2081/92.
[145] Regulation (EU) No 1151/2012 of the European Parliament and of the Council of 21 November 2012 on quality schemes for agricultural products and foodstuffs.
[146] *Cf.* Single document of the PDO 'Parmigiano Reggiano' – OJ C 87/16, 16.04.2009.
The other EU agricultural product quality policy measures concerning GIs protection are laid down in the following regulations:
- Council Regulation (EEC) No 1601/91 of 10 June 1991 laying down general rules on the definition, description and presentation of aromatized wines, aromatized wine-based drinks and aromatized wine-product cocktails;
- Council Regulation (EC) No 1234/2007 of 22 October 2007 establishing a common organisation of agricultural markets and on specific provisions for certain agricultural products (Single CMO Regulation) and in particular in Part II, Title II, Chapter I, Section I and in Section Ia, Subsection I thereof;
- Regulation (EC) No 110/2008 of the European Parliament and of the Council of 15 January 2008 on the definition, description, presentation, labelling and the protection of geographical indications of spirit drinks.

Under EU legislation, in particular as concerns agricultural products and foodstuffs Regulation (EU) No 1151/2012, the registration of a PDO requires that the agricultural product or foodstuff complies with a product specification: according to Article 7 the product specification of a PDO shall include, for example, the description of the agricultural product or foodstuff, including the raw materials and the main physical, chemical, microbiological or organoleptical characteristics, the definition of the geographical area, evidence that the product originates in the defined geographical area (the above mentioned so-called area of origin), the description of the method for obtaining the product, details bearing out the link between the quality or the characteristics of the product and the geographical environment, the name and address of the authorities or bodies verifying compliance with the specification, and any specific labelling rule.

According to the case-law of the Court of Justice of the European Union, the product specification:

> determines both the extent of the obligations to be complied with for the purposes of using the PDO and, as a corollary, the extent of the right protected against third parties by the effect of registration of the PDO, which lays down at community level the rules set out or referred to in the specification.[147]

Under EU legislation, official controls cover verification that a product complies with the corresponding product specification and such controls are carried out, in accordance with Regulation (EC) No 882/2004, by competent authority or authorities designated by the Member State and/or by one or more of the control bodies within the meaning of point (5) of Article 2 of Regulation (EC) No 882/2004 operating as a product certification body.

5.2.3 International framework

With respect to the international safeguard, several multilateral agreements provide different degrees of protection to Intellectual Property, including Geographical Indications (GIs).

The Paris Convention for the Protection of Industrial Property, of 20 March 1883 is the oldest of these IP international agreements (Article 1(2) of the Convention), including appellations of origin, even if the Convention does not grant any definition of them.

The Stresa Convention on the Use of Appellations of Origin and Denominations of Cheeses of 1 June 1951 is the oldest specific international agreement concerning the protection of designations of origin, even if related to only a specific type of

[147] Case C-469/00, judgment of 20 May 2003, point 80.

foodstuffs, *i.e.* cheeses. This agreement is still in force for some members, such as Italy and Switzerland[148]. The name Parmigiano Reggiano is in the Attachment A, concerning the appellations of origin.

In 1958 the first international agreement for the protection of the designation of origin for all goods came to light, *i.e.* the Lisbon Agreement for the Protection of Appellations of Origin and their International Registration of 31 October 1958[149], managed by WIPO. To date 28 countries have joined it, among which Italy, France, Mexico and Cuba: as regards the registration of the Appellation of Origin Parmigiano Reggiano, it dates back to 23 December 1969.

However, the most recent and probably important international step for GIs protection is the Agreement on Trade Related Aspects of Intellectual Property Rights (TRIPs) of 15 April 1994, which addresses obligations to all WTO members for the IP protection, including GIs in Section 3 of its Part II (Articles 22 to 24)[150]. However, it is commonly known that the TRIPs Agreement provides for two different levels of protection: the minimum standards of protection, concerning the deceptiveness for the public and the unfair completion for GIs, for all products; and the additional protection only for GIs for wines and spirits, which goes beyond any risk of confusion.

Moreover, in addition to international agreements, GIs may be protected by Free Trade Agreements or bilateral agreements. To this end, the European Union (EU) has recently concluded several agreements with third countries or regions: just as examples, Switzerland, South Korea, and Central America.

Finally, a GI may be registered as such in countries that have established a *sui generis* system of protection: to this end, the name Parmigiano Reggiano has already been registered as designation of origin in many third countries (such as Albania, Belarus, Mongolia, Norway, Russia, and Ukraine).

5.3 The 'Consorzio del Formaggio Parmigiano-Reggiano'

The Consorzio del Formaggio Parmigiano-Reggiano[151], created in 1934, is a voluntary not-for-profit body established in accordance with Italian law which associates all the Parmigiano Reggiano cheese producers.

---

148 Court of Milan, Industrial and Intellectual Property Special Section, 17 March 2012, 3196/12.
149 Lisbon Agreement for the Protection of Appellations of Origin and their International Registration of October 31, 1958, as revised at Stockholm on July 14, 1967, and as amended on September 28, 1979. www.wipo.int/treaties/en/text.jsp?file_id=285856.
150 Agreement on Trade-Related Aspects of Intellectual Property Rights www.wto.org/english/tratop_e/trips_e/t_agm0_e.htm.
151 Cf. the website www.parmigianoreggiano.com/default.aspx.

It is a group under Articles 2 and 45[152] of Regulation (EU) 1151/2012 and is recognised by the Italian Ministry of agriculture, food and forestry for the protection, safeguard, promotion, consumer information and general management of the PDO Parmigiano Reggiano.

All the following operators, provided that they are located in the area of origin and subject to the official controls for the PDO Parmigiano Reggiano, may become members of the Consorzio: *i.* farmers producing milk for the production of Parmigiano Reggiano; *ii.* dairies producing cheese suitable to become Parmigiano Reggiano (presently, 100% is member of the Consorzio); *iii.* operators maturing cheese suitable to become Parmigiano Reggiano and holding the product till the end of the minimum maturation period and operators engaged in the portioning and/or grating and packaging of cheese bearing the PDO Parmigiano Reggiano.

The Consorzio is the custodian, the guardian of the PDO Parmigiano Reggiano in the interest of both Parmigiano Reggiano producers and consumers. In particular, the functions of the Consorzio, also cooperating with the Ministry of agriculture, food and forestry policies, are aimed at the supervision, protection and safeguard of the PDO Parmigiano Reggiano from any form of abuse, counterfeiting, illegal and improper use of the protected designation of origin.

In Italy, its supervisory agents bear the qualification of policemen whereas, in the EU territory, the Consorzio avails of the assistance of the national competent authorities on the marketplace designated under the Articles 13(3), 36(3) letter b) and 38 of Regulation (EU) No 1151/2012 (the so-called *ex-officio* protection), and Articles 54 and 55 of Regulation (EC) 882/2004 to prevent or stop the unlawful use of a PDO Parmigiano Reggiano in accordance with the scope of protection established in Article 13 of Regulation (EU) No 1151/2012.

Furthermore, the Consorzio is owner of several collective/certification trademarks, which are used for distinguishing the cheese bearing the PDO Parmigiano Reggiano, *i.e.* the cheese that has been certified in compliance with the specification by the relevant control body[153].

[152] Role of groups (Article 45) – 'A group is entitled to, inter alia, take action to ensure adequate legal protection of the protected designation of origin or protected geographical indication and of the intellectual property rights that are directly connected with them' (Article 45(1)(a)).

[153] For the PDO Parmigiano Reggiano, the control body under Articles 36 and 37 of Regulation (EU) 1151/12 is 'Organismo Controllo Qualità Produzioni Regolamentate Soc. Coop'. www.ocqpr.it/ocq-pr/presentazione.html.

As regards the milk products sector, it is further important to highlight that Regulation (EU) No 261/2012 (the so-called 'Milk package')[154], which entered into force on 2 April 2012, introduced the 'Regulation of supply for cheese with a protected designation of origin or protected geographical indication', applicable from 3 October 2012[155].

As concerns the promotion of the PDO Parmigiano Reggiano, the Consorzio carries out activities, collaborates with sportsmen, sportswomen, chefs, participates to food fairs, organises shows and tours to the dairies, etc.: the promotion activity concerns a very important part of the budget.

5.4 The enforcement of the PDO Parmigiano Reggiano within the European Union

Within the European union, GIs benefit from a strong system of protection under several aspects:
- scope of protection;
- enforcement.

5.4.1 Scope of protection

In accordance with EU legislation and case-law of the Court of Justice of the European Union and of the national Courts, GIs enjoy a wide scope of protection.

One of the most interesting and innovative aspects of the wide protection granted to GIs within the EU is undoubtedly the 'evocation', on which the Court of justice has repeatedly focused, clearly outlining the meaning.

Firstly, the Court of Justice, in case C-87/97, judgment 4 March 1999, pointed out that:

> it is possible for a protected designation to be evoked where there is no likelihood of confusion between the products concerned and even where no Community protection extends to the parts of that designation which are echoed in the term or terms at issue[156].

[154] Regulation (EU) No 261/2012 of the European Parliament and of the Council of 14 March 2012 amending Council Regulation (EC) No 1234/2007 as regards contractual relations in the milk and milk products sector OJ L 94, 30.3.2012.
[155] Regulation (No) 261/12, whereas (17) (24) and Article 126d.
[156] Cf. C-87/97, Judgment 4 March 1999, Paragraph 26.
Consorzio per la tutela del formaggio Gorgonzola vs Käserei Champignon Hofmeister GmbH & Co. KG and Eduard Bracharz GmbH.

In the same way, in a case concerning an action brought by the European Commission against the Federal Republic of Germany, – the Commission assuming the obligation for Germany to protect *ex officio* the PDOs/PGIs under the first Regulation (EEC) 2081/92 -, the Court of justice (Grand Chamber), judgment 26 February 2008, recalling the previous decision, stated that:

> proximity and the phonetic and visual similarities referred to in Paragraph 46 above are such as to bring to the mind of the consumer the cheese protected by the PDO 'Parmigiano Reggiano', when he is confronted by a hard cheese, grated or intended to be grated, bearing the name 'Parmesan'.[157]

Lastly, in a case concerning a conflict between a trademark and the GI 'Cognac' under Regulation (EC) 110/08, the Court pointed out that:

> Points (a) to (d) of Article 16 of Regulation No 110/2008 refer to various situations in which the marketing of a product is accompanied by an explicit or implicit reference to a geographic indication in circumstances liable to mislead the public as to the origin of the product or, at the very least, to set in train in the mind of the public an association of ideas regarding that origin, or to enable the trader to take unfair advantage of the reputation of the geographical indication concerned[158].

In accordance with the principles established by the Court of Justice, in Germany, the Court of Appeal of Berlin, judgment 15 June 2010[159], upheld the judgment of the Court of Berlin of 22 April 2008 where the Court, in line with the scope of protection provided for in Article 13(1)(b) of Regulation (EC) 510/06, in an action brought by the Consorzio, ordered a German operator to refrain from producing, advertising and offering for sale on the German market cheese under the names 'Parmesan' and 'Parmigiano' if not complying with the specification for the PDO Parmigiano Reggiano.

Again in Germany, the Court of Cologne, orders of 14 October 2009 and 12 October 2011, in preliminary injunctions actions brought by the Consorzio, ordered US operators to refrain from advertising in Germany cheeses under the denominations 'Parmesan', 'Parmesano' and/or 'Reggianito' not complying with the PDO 'Parmigiano Reggiano' specification, exhibited at Anuga fair[160].

---

[157] Cf. C-132/05, Judgment 26 February 2008, Commission of the European Communities vs Federal Republic of Germany, Paragraphs 48/49.

[158] Cf. Joined Cases C 4/10 and C 27/10, Judgment 11 July 2011, Bureau national interprofessionnel du Cognac vs Gust. Ranin Oy.

[159] Court of Appeal of Berlin, judgment 15 June 2010 in case 5 U 97/08.

[160] Court of Cologne, order of 14 October 2009, case 31 O 610/09.
Court of Cologne, order of 12 October 2011, case 31 O 630/11.

Whereas in Spain, in accordance with the consolidated principles expressed by the Court of justice, the Commercial Court of Oviedo considered as well the use of the denomination 'Parmeso' for grated cheese an evocation of the PDO 'Parmigiano Reggiano'. It should be noted that this case is of particular significance because the Court confirms that the Consorzio had standing not only to act in defence of the PDO, but also to seek damages, including on behalf of its members, particularly given the risk that the latter would not bring individual actions in a foreign forum (in particular where it concerns a limited amount of damages) or the safeguard of their rights[161].

Finally, it is interesting to note that, unlike as for the trademarks scope of protection[162], the EU GI's regulations do not provide for any limitation of the effects of the protection.

The PDO Parmigiano Reggiano enforcement has also been implemented through administrative actions: following a cancellation action of the Consorzio, the OHIM Cancellation Division, decision of 8 July 2013, cancelled partially the registered community trademark (figurative) 'TRADIZIONI DI PARMA' for milk products and cheese[163].

5.4.2 Enforcement

In addition to the possibility for the producers and the group of producers, such as the Consorzio, to bring actions before the Courts (private enforcement), the new Regulation (EU) 1151/12 has finally clearly introduced the so-called '*ex officio*' protection[164].

According to the obligations provided in Articles 13(3), 36(3)(b) and 38 of Regulation (EU) 1151/12, and Articles 54 and 55 of Regulation (EC) 882/04, it may be reasonably argued that presently the Member States' authorities shall carry out official controls for the monitoring of the use of PDOs/PGIs on the market and, in the event

[161] Commercial Court of Oviedo, judgment n. 65/2010 of 12 January/ 5 April 2010.

[162] Cf. Council Regulation (EC) No 207/2009 of 26 February 2009 on the Community trade mark, Article 12.

Court of Berlin, order of 23 August 2005, 102 O 60/05.

[163] Cancellation Decision 6890 C of 8 July 2013. Appeal decision is pending.

[164] Actually, the Court of justice, in the case concerning an action brought by the European Commission against the Federal Republic of Germany, assuming the Commission the obligation for Germany to protect ex officio the PDOs/PGIs under the first Regulation (EEC) 2081/92 referred to above, stated that '*It must be held that the Commission has not established that, by formally refusing to proceed against the use on its territory of the name "Parmesan" on the labeling of products which do not comply with the requirements of the specification for the PDO 'Parmigiano Reggiano', the Federal Republic of Germany has failed to fulfill its obligations under Article 13(1)(b) of Council Regulation (EEC) No 2081/92*'.

Cf. C-132/05, Judgment 26 February 2008, Commission of the European Communities vs. Federal Republic of Germany, Paragraph 81, cited above.

of breaches, take appropriate steps to prevent and/or stop the unlawful use of protected designations of origin and protected geographical indications that are produced or marketed in that Member State[165].

5.5 The enforcement of the PDO Parmigiano Reggiano outside of the European Union

As regards the GIs protection outside of the EU, the situation may differ from country to country.

First of all it has already been noted that the TRIPs Agreement, which addresses obligations to all WTO members for the IP rights protection, including GIs (Article 22 to 24), provide for two different levels of GIs protection: the minimum standards of protection, prohibiting the deceptiveness for the public and the unfair competition, for GIs for all products; the additional protection only for GIs for wines and spirits, beyond any risk of confusion.

The EU GI's legislator, as mentioned above, has gone beyond the obligations arising from TRIPs, as it has provided a high level of protection, beyond any risk of confusion[166], not only for GIs for wines and spirits, but also for agricultural products and foodstuffs[167].

As a consequence, the GIs protection may differ from country to country: for GIs' for products other than wines and spirits, a GIs protection against deceptiveness for the public and unfair competition would lead a WTO member country to comply with the TRIPs obligations.

In light of the above, in addition to the 'old generation' specific agreements concluded by the EU with third countries on the protection of GIs' for wines and spirits, the EU has recently sought to get a high protection in third countries also for GI's for agricultural products and foodstuffs, including the GIs protection in the negotiation of FTA (Free Trade Agreements) or through specific agreements on the GI's protection[168].

---

[165] The names and addresses of the Member States authorities are published on the web site of the Commission http://ec.europa.eu/agriculture/quality/schemes/compliance-authorities_en.pdf.

[166] And even higher than the additions protection provided for in TRIPs Agreement for GIs for wines and spirits.

[167] In July 2012 the European Commission has published a Study on Geographical Indications (GIs) for non- agricultural products in the European Union (EU).
Cf. *oriGIn Flash*, 26 March 2013 (organization for an International Geographical Indications Network – oriGIn) www.origin-gi.com/. The Study is available at http://ec.europa.eu/internal_market/indprop/geo-indications/index_en.htm.

[168] O'Connor, Richardson, The legal protection of Geographical Indications in the EU' Bilateral Trade Agreements, moving beyond TRIPs, Rivista di diritto alimentare, Anno VI, numero 4, Ottobre-Dicembre 2012 http://rivistadirittoalimentare.it.

To this end, just as examples, the extent of protection provided for in specific agreement on the protection of designations of origin and geographical indications for agricultural products and foodstuffs, such as the one concluded with Switzerland[169], or in FTAs, such as the first one including the GIs protection concluded with South Korea[170], is high.

A further very interesting legal tool for the GI protection is the Lisbon Agreement for the Protection of Appellations of Origin and their International Registration of 31 October 1958[171], managed by WIPO, which presently gathers 28 member states.

As mentioned before, the denomination Parmigiano Reggiano was registered within the Lisbon System on 23 December 1969, and no declaration of refusal was presented.

Finally, as referred to above, a GI may be registered as such in countries that have established a *sui generis* system of protection: to this end, the name Parmigiano Reggiano has already been registered as designation of origin in many third countries (including among others Albania, Belarus, Mongolia, Norway, Russia, and Ukraine). However, it does not mean that any *sui generis* system provides for a high level of protection for GIs for products other than wines and spirits: as mentioned before, even a *sui generis* system for GIs' protection for such products, granting protection against deceptiveness for the public and unfair competition, would lead a WTO member country to comply with its TRIPs obligations.

[169] Agreement between the European Union and the Swiss Confederation on the protection of designations of origin and geographical indications for agricultural products and foodstuffs, amending the Agreement between the European Community and the Swiss Confederation on trade in agricultural products http://eur-lex.europa.eu/LexUriServ/LexUriServ.do?uri=OJ:L:2011:297:0003:0047:en:PDF.

[170] Free Trade Agreement between the European Union and its Member States, of the one part, and the Republic of Korea, of the other part http://eur-lex.europa.eu/LexUriServ/LexUriServ.do?uri=OJ:L:2011:127:0006:1343:en:PDF.

[171] Cited above.

6. Infringement of protected geographical designations by means of evocative behaviours – a critical prospective

Ángel Martínez Gutiérrez

6.1 Preview

European Institutions started in the early nineties to provide some harmonised legal protection to qualified geographical indications. In this system of protection of product names, emphasis must be given to the recognition of an exclusive right that, in general, is parallel to the right granted by a trade mark registration. This regulation implies an advance which leaves behind the protection dispensed only and exclusively through the mechanisms protecting against unfair competition. It provides the group of businessmen who have promoted and obtained the official recognition of the protected designation of origin (PDO) or protected geographical indication (PGI) with an opportunity to control the future use of the specific geographical name to identify a product. To that end, they are provided with a number of powers that allow them to react against any unlawful use of the same name. In this way, the two main functions of titles provided by Intellectual Property Law are guaranteed, namely: the ability to differentiate products with specific provenance from others of different geographical origin and to distinguish a productive quality linked to their provenance, which facilitates the recognition and rational choice by the consumers.

Despite these assertions, if we carry out an in-depth comparison of the rights granted by the recognition of a protected geographical indication and those granted with the registration of a trade mark, we will see that the exclusivity rights which are recognised in both cases do not enjoy either the same extension or the same level of protection. For this reason we have used the word *parallel* to define how the systems of trade mark and protected geographical indications are configured, as we consider that beyond the name and structure itself, it is not possible to maintain a comparison between the two types of regulations. Thus, from a strictly formal level, the regulation of exclusivity rights in the case of the trade mark is of a technical nature which contains a clear definition of its essential features, whereas the regulation of exclusivity rights in protected geographical indications is, in any case, more precarious. This legal disparity justifies that in the field of protected geographical indications there is a permanent resort to the interpretative criteria used in the case of trade marks, which enjoy better legal regulation and dispose of a great number of judicial decisions that contribute to interpreting the different legal provisions. In this way, the shortcomings of the specific legal regime can be overcome and some help is provided to guide the interpretative work of the administrative or judicial entities that have to apply it. Unfortunately, as we shall see in this chapter, the transfer of these interpretative criteria is not complete and we will arrive at solutions that break the legal assumptions that have traditionally been maintained in the field of trade marks.

Moreover, this regulatory situation of bigger and better protection offered in the field of trademarks provokes that, in practice, the name of the protected geographical designation should also be registered as a trade mark in order to obtain better legal protection. Two reasons support this option. Firstly, because the protective instruments of the trade mark are added to the specific protection of the geographical designation. Indeed, from this moment the structure of development, control and defence of the protected geographical designation may use either its specific powers of exclusion or those of the protected trade mark. Secondly, the practice of judicial decisions shows that administrative and judicial bodies which adjudicate on the matter are more familiar with Trade Mark Law, a system of Law that they know in more depth and are better prepared to protect.

In these pages we will address one of the issues which arise out of the legal status of protected geographical indications, namely, the exegesis of the term 'evocation', a concept on which is built one of the autonomous conducts prohibited by the Law. In order to do so we will make a brief presentation of the legal protection of recognised geographical indications and subsequently we will examine the European legal provision defining the presence of the *ius excludendi alios* (the right to exclude others) in these collective intellectual property titles. After this, we will carry out the analysis of the concept of 'evocation' confronting the interpretation offered by Spanish courts and the European Court of Justice (ECJ). This comparative approach will allow us to test which of the two systems provides a better tutelage(guardianship) of these protected geographical indications.

6.2 Introductory remarks on community tutelage of protected geographical indications

Geographical indications are specially and strongly protected under European Union Law, a protection which has received a new impulse with Regulation (EU) 1151/2012 of 21 November 2012, on quality schemes for agricultural products and foodstuffs, a Regulation which improves and repeals the former Regulation (EC) 509/2006 on agricultural products and foodstuffs as traditional specialities guaranteed and Regulation (EC) 510/2006 on the protection of geographical indications and designations of origin for agricultural products and foodstuffs. The protection provided by the new Regulation is based on the recognition of an exclusivity right for a group of legitimate users, provided that a valid registration has been made in the Register created for this purpose by the European Commission – safeguarding the transitional national protection -. Similarly to what happens in the field of trade marks, the exclusivity right is obtained by the registration and benefits the collective that has driven (and obtained) the registration of geographical indication. However, only in the case of a trade mark is the right exclusively granted to the applicant. This exclusivity is twofold, positive and negative, and it is orientated towards the achievement of two trade functions that we have mentioned before, namely, a distinctive function and a quality assurance function. Therefore, while the positive side (*ius utendi*) or monopoly of use allows the effectiveness of both functions in the course of trade, the negative side (*ius excludendi alios*), or powers of exclusion, aims at preventing some unfair behaviours of third parties, within and outside the collective group, in the market.

Since the *ius excludendi alios* applying to protected geographical names will be analysed in the next section, we believe we should make some observations first on the positive side of this right which is included in the first Paragraph of Article 12 of Regulation (EC) 1151/2012. According to this Article 'protected designations of origin and protected geographical indications may be used by any operator marketing a product conforming to the corresponding specification'.

Nevertheless, even if the recognition of the *ius utendi* has been made in very generous terms, the attentive reading of the transcribed article suggests a double parameter which restricts to some extent the initial amplitude of the rule. We should note, in this regard, that by recognising a monopoly of use for 'any operator marketing a product' this rule would be incomprehensibly vetoing the use of the geographical name to those traders who develop their business in other stages of the productive process (and in the earlier phase, in all cases, of the indicated stage). Such is the case of production, processing and preparation. Therefore, there is an unjustified restriction to the field of application of this rule which, of course, is not coherent with the defining characteristics of protected geographical names. It is for this reason that, in our opinion, this unfortunate fault should be overcome by means of the systematic interpretation of that rule with the definitions of the titles of protection that appear in the first two Paragraphs of Article 5 of Regulation (EC) 1151/2012. Thus, the proposed exegesis would try to complete this deficiency by extending the verb 'marketing' to all the stages of production that are referred to in the specification for the particular geographical designation.

Along with these considerations, the reading of the rule reveals that the exclusive use of a registered geographical indication is made around an entirely justified condition, namely, the compliance with the requirements of the particular specifications for production, development and commercialisation of differentiated products having a specific geographical name. And, by requiring that agricultural or food products are conforming to the specifications applicable to them', the Union Regulation pursues that geographical names obey the two functions listed above and which, in our view, constitute an adapted version from those developed for the distinctive signs[172] in the market. We should note in this regard, how the monopoly of use recognised to legitimate users of the geographical name facilitates that this name becomes a differentiating instrument of collective character which allows us to identify and distinguish products from specific geographical origin, whose characteristics appear homogeneous and permanently certified over time.

6.3 Approach to the negative dimension of the exclusivity right

6.3.1 Drafting of the rule, a critical remark

Together with the monopoly of use recognised to a more or less broad group of persons, the exclusivity right of protected geographical designations also implies the attribution of a power of exclusion or *ius excludendi alios*, to implement effectively the said monopoly and to grant the legal and economic functions of this title of Intellectual Property Law. The regulation of this *ius prohibendi* is included in Articles 13(1) and 14 of Regulation (EC) 1151/2012.

These two articles are certainly complex and provide the protected geographical names with a broad and solid *ius prohibendi*. Therefore, on the one hand, the task of facilitating the inspection, promotion and defence of the geographical name is fulfilled by vetoing certain harmful behaviours of third parties, who may or may not belong to the circle of persons entitled to use the geographical name. On the other hand, these articles help to solve disputes with other distinctive signs, by ruling on the differences that apply between them. In fact, if we return to the content of both legal rules, we can detect the design provided by the Law for the defence of protected geographical indication. While Article 13, entitled 'protection', makes a positive definition (and a negative one) of the behaviours susceptible to misuse a particular geographical registered name, Article 14, entitled 'relations between trade marks, designations of origin and geographical indications', is devoted to rules on conflicts between both titles of Intellectual Property Law, establishing a kind of graduation of the protection accorded to geographical indications in each case. [173]

For the purposes of this chapter, it is essential to focus on the content of the first Paragraph of Article 13 of Regulation (EU) 1151/2012 because it contains a positive definition of the prohibited conducts whose infringement generates the activation of the powers provided to the group of legitimate users of the protected geographical

[172] The term 'sign' is used in this paper to represent any sign, mark, symbol, name, denomination or other representation of the qualities of a product or service, as appropriate in the relevant context.
[173] For an analysis of the game of both provisions, you can consult our works, A. Martinez Gutiérrez, 'La tutela comunitaria de las denominaciones geográficas protegidas ante las marcas registradas', *NotUE*, No. 219, 2003, pp. 27 y ss.; and also the monograph '*Tutela comunitaria de las denominaciones geográficas…*', cit., 52 y ss.

indication[174]. The current wording of Regulation (EU) No. 1151/2012 involves some technical improvements to the wording of the repealed Article 13 of Regulation (EC) 510/2006, which in turn implied a minimal qualitative improvement over the original version included in Regulation (EC) 2081/1992. Nevertheless, as far as the content of these regulations is concerned we can conclude that the actual scope of the rule has remained unchanged after the reforms, a fact which allows us to resort to what was expressed both by the scientific doctrine and by the case law concerning the former regulations. In this sense, from a strict legal-technical perspective, the first Paragraph of Article 13 of Regulation (EU) 1151/2012 expresses, in our view, an unlucky character, because pretending to give the protected geographical names a wide *ius prohibendi*, provides a confusing and repetitive description of the specific infringements and also omits a necessary tribute to legal concepts developed in the field of distinctive signs, which, in our opinion, is extremely serious if we remember that at the time of the drafting and adoption of the mentioned Regulation, the legal regime of the brands was an important matter of Community character[175].

Therefore, by accepting and reproducing the generous formulation of the factual premises of the preceding rules, Community legislation facilitates the application of the present legal provision to a wide range of behaviours capable of infringing the protected geographical names. In this context, we can remark how, on the one hand, 'Any direct or indirect commercial use', 'any other false or misleading indication' or 'any other practice' capable of infringing the exclusive right of a geographical designation are prohibited while, on the other hand, prevention, as the legislative politics option, is chosen by activating the protective control in a stage previous to the effective harmful result. In this way, Paragraphs (c) and (d) of Article 13(1) come to veto respectively not only false or misleading indications

[174] Article 13(1) establishes that 'Registered names shall be protected against:
(a) any direct or indirect commercial use of a registered name in respect of products not covered by the registration where those products are comparable to the products registered under that name or where using the name exploits the reputation of the protected name, including when those products are used as an ingredient;
(b) any misuse, imitation or evocation, even if the true origin of the products or services is indicated or if the protected name is translated or accompanied by an expression such as 'style', 'type', 'method', 'as produced in', 'imitation' or similar, including when those products are used as an ingredient;
(c) any other false or misleading indication as to the provenance, origin, nature or essential qualities of the product that is used on the inner or outer packaging, advertising material or documents relating to the product concerned, and the packing of the product in a container liable to convey a false impression as to its origin;
(d) any other practice liable to mislead the consumer as to the true origin of the product.'

[175] See: D. Sarti, 'Regolamento CE n. 510/06 sulle denominazioni d´origine' en *Commentario breve...*, cit., p. 1035, quien defiende el recurso de los criterios y parámetros utilizados en el ámbito de los signos distintivos; S. Magelli, 'Marchio e nome geografico' en AA.VV., *Studi di Diritto Industriale in onore di Adriano Vanzetti. Proprietà intellettuale e concorrenza*, vols. I y II, Milán, 2004, pp. 916 y 917, quien afirma que la tutela de las denominaciones geográficas protegidas *'in realtà è una tutela ben poco definita'* y que los derechos reconocidos a aquéllas son *'diritti più sfumati'*.

'liable to convey a false impression as to its origin', but also 'any other practice liable to mislead consumers about the true origin of the product', regardless of whether a certain and effective situation of confusion is created in the market.[176]

However, even though this positive effect on the protection of geographical names registered is achieved, the article does not make a clear definition of harmful behaviours as is the case in the field of distinctive signs, where the exercise of the powers of exclusion by the trade mark holder is conditional upon the concurrence of specific factual circumstances that are expressly described in the rule. On the contrary, the configuration of the factual premises reveals the poor legal technique invested in the development in the first paragraph of Article 13 of Regulation (EU) 1151/2012, because it is activated when any of the behaviours written in a vague and undetermined way by the Community legislation is committed. We must add that these are behaviours that usually provoke the risk of confusion, to take unfair advantage of the reputation of others or risk misleading about some characteristics of the product.

6.3.2 Forbidden behaviours

The first paragraph of Article 13 of Regulation (EU) No. 1151/2012 prohibits a range of behaviours in order to avoid certain negative effects in the market. In its analysis we will take advantage of the above-mentioned connection of this sector with the trade mark field, a connection which will allow us to resort to the evaluative parameters used by doctrine and Spanish and comparative case law in this field.[177]

In the first place, the risk of confusion is one – the first one – of the negative consequences which the mechanics of Article 13 of Regulation (EU) No. 1151/2012 try to avoid. Indeed, although this is not expressly referred to, we think it is set out in the first paragraph, in sub-paragraph (a), of this Article. This is a relevant provision if we bear in mind that one of the functions of geographical designations in the market is the distinctive one, mitigating the negative effects of the informative asymmetry existing within it and, in consequence, allowing consumers an easy choice of products by means of clear and concise data about the origin. It is evident that the first effect of the Community legislation is targeted at vetoing

---

[176] These circumstances have come to substantiate a thesis about the extent of *ius excludendi alios* as it applies to recognized protected geographical designations. See M.M. Gomez Lozano, *Denominaciones de origen y otras indicaciones geográficas*, Cizur Menor, 2004, pp. 94 y ss.; M. Botana Agra, *Las denominaciones de origen*, Madrid, 2001, pp. 204 y ss.

[177] This is shown in: M. Botana Agra, *Las denominaciones de origen*, cit., pp. 207 y 208, who affirms in relation to the interpretation of the concept of imitation in the Regulation (EU) No. 2081/92 that '... will have to go to any rules or guidelines that Member States apply in resolving disputes concerning the existence or not of confusing similarity of distinctive signs, particularly those consecrated in the sector of private-label...'.

the unfair behaviours of those trying to vitiate the act of purchasing and provoke a situation that is close to avoidance of the use of distinctive signs in general and of geographical names in particular.[178] Hence the EU legislator has adopted this provision that makes the distinctive function fundamental and ensures its effectiveness.

Nevertheless, if we focus on the analysis of this subsection and contrast it with the EU Regulation on trade marks, we can verify that this provision is defective[179]. Therefore, when prohibiting 'any direct or indirect commercial use of a registered name in respect of products not covered by the registration where those products are comparable to the products registered under that name', this article is making an implicit but defective reference to the parameters upon which the risk of confusion in the field of trade mark are established. Indeed, by omitting any explicit reference to the mentioned negative effect whose concurrence is incontestable, the factual premises require a determined structure on the side of the infringing sign ('direct or indirect commercial use of a registered name') and a proximity between products conflictingly offered under similar denominations, ('products not covered by the registration where those products are comparable to the products registered'). These are definitely two parameters which – as Galli has graphically affirmed in the trade mark field – could be represented on the axes of ordinates and abscissas of an assumption diagram, where all possibilities of confusion for the public consumer are represented[180], and among which there exists an undisputed interdependence leading to a risk of confusion, since as stated by the Court of Justice of the European Communities 'a lesser degree of similarity between those products or services may be offset by a greater degree of similarity between the brands and *vice versa*'.[181]

Secondly, the first paragraph of Article 13 of Regulation (EU) 1151/2012 condemns those behaviours capable of unduly exploiting the reputation of the protected geographical name.

Indeed, more explicitly, the EU legislation refers to this negative effect in sub-Paragraphs (a) and (b) of the mentioned Article. Sub-paragraph (a) prohibits:

[178] In this way is pronounced the considered Article 5 of Regulation 1151/2012, when considering that the rules of regulation are arranged 'not only to identify and promote better those of its products with specific characteristics, but also to protect mentioned producers from unfair practices'. For this purpose, adds the considered article 18 that is necessary 'provide clear information on products with specific characteristics linked to geographical origin, for consumers to make their purchasing choices with more informed'.

[179] A. Martinez Gutiérrez, *Tutela comunitaria...*, cit., pp. 81 y ss.

[180] C. Galli, *Funzione del marchio e ampiezza della tutela*, Milán, 1996, p. 105.

[181] Paragraphs No. 17 and 19 of the Judgments of the Court of Justice of 29 September, 1998 (case C-39/97, *Canon*) and of 22 June, 1999 (case C-342/97, *Lloy*), respectively.

> any direct or indirect commercial use of a registered name in respect of products not covered by the registration, where...the use of the name exploits the reputation of the protected name, including when those products are used as an ingredient.

Sub-paragraph (b) refers, generally, to a trilogy of cases where the effect is to exploit another's reputation, namely:

> any misuse, imitation or evocation, even if the true origin of the products or services is indicated or if the protected name is translated or accompanied by an expression such as 'style', 'type', 'method', 'as produced in', 'imitation' or similar, including when those products are used as an ingredient.[182]

From the joint analysis of both rules it is possible to appreciate that the regulation admits that distinctive signs can have different configurations, being able to present – applying the criteria of the distinctive signs' own hermeneutic – a graphical, phonetic and/or conceptual similarity to the geographical protected name.

The Union legislation on protected geographical designations also represents a notable advance regarding the regulation of the risk of confusion and the improper use of reputation. Therefore, in relation to the production area where the activity takes place, the comparable characteristics of the products are not required to satisfy the specification covered by the registration of the geographical name[183]. In line with legislation on trade marks, the use of another's reputation goes beyond the principle of speciality, this rule being applicable anyway to the case of non-similar products. In this sense, the Articles 4(4)(a) of the Trade Mark Directive[184] and 8(5) Community Trade Mark Regulation,[185] implemented into Spain by Article 8(1) of Act 17/2001 of 7 December on Trade Marks, where distinctive marks enjoying notoriety and well known in the market are protected when confronted by identical or similar marks in order to differentiate similar products or services

[182] It is interesting to contrast the wording of the current provision with that of Article 13 of the repealed Regulation No. 510/2006, with that mentioned between examples of prohibited keywords: gender, type, method, style and imitation. While the meaning of the rule is unchanged, it is difficult to understand why the change in terms.

[183] So says, C. Galli, 'Globalizzazione dell´economia e tutela della denominazioni di origine dei prodotti agro-alimentari', *Riv. Dir. Ind.*, 2004, I, p. 71, who believes that Article 13(1)(b) of Regulation (EU) No. 2081/1992 grants protection beyond the principle of specialty, confirming an approximation between the trade marks and protected geographical indications. For his part, L. Sordelli, 'Denominazione di origine ed indicazioni geografiche', *Enciclopedia del Diritto*, Aggiornamento I, Milán, 1997, pp. 501 y 502, argues that the geographical name is also protected against any use of reputation from the use of signs in the field of comparable products.

[184] See Directive 2008/95/EC of the European Parliament and of the Council of 22 October 2008 to approximate the laws of Member States relating to trade marks (codified version).

[185] See Regulation (EC) No. 207/2009 of the Council of 26 February 2009 on the Community trade mark (codified version).

which can improperly take advantage of the reputation of a protected product or damage its distinctive nature.[186]

Finally, the legislation of the European Union expresses a paradigmatic triple assumption of unfair misappropriation of reputation,. despite any inclusion of indications of the true origin or a translation of the geographical designation,[187] or if that:

> is accompanied by expressions such as 'style', 'type', 'method', produced as in 'imitation' or similar expressions.

This is a prohibition that, of course, recalls the second paragraph of Article 12 of Spanish Act 3/1991 of 10 January on Unfair Competition, which states that:

> deemed as unfair competition are the use of distinctive signs belonging to another person or of false designations of origin accompanied by the indication of the true origin of the product or expressions such as 'model', 'system', 'type', 'class' and the like'.[188]

Similarly, this Community principle echoes also the standard of protection of geographical indications in the wine-making field. In this sense, it can be corroborated in Articles 23(1) of the TRIPs Agreement and 5(1)(h) of the Spanish Trade Marks Law, which respectively prohibit the use of geographical indications to identify wines or spirits not having this origin:

> even when the true origin of the product is specified or the geographical indication translated or accompanied by expressions such as class, type, style, imitation or similar is used[189].

---

[186] See, C. Fernández -Novoa, Trade mark Law Treaty, 2nd ed., Madrid, 2004, p. 389 et seq. E. Bastian, 'Il marchio celebrate European Diritto nel nella dei marchi e della Corte di giurisprudenza European Giustizia' in Studi di Diritto Industriale ..., vol. I, cit., Pp. 109 et seq.

[187] Even when it involves an obvious advantage of the reputation achieved by the specific geographical designation, the inclusion of dislocative clauses comes to eliminate the risk of error in the consumer. On this important function, consult, inter alia, C. Fernández -Novoa, International protection of geographical names of products, Madrid, 1970, p. 34 et seq. M. Lalín Arean, 'Definition and legal protection of geographical indications ', ADI, t. XIV, 1991-1992, p. 69; G. Buccirossi, 'The American stile dei jeans' Il Diritto Industriale, 1994, pp. 398 and 399. Contrary Marteau – Roujou OF Boubée, I., Les marques deceptives, Droit français, Droit communautaire, Comparative Law, Paris, 1992, pp. positions. 87 et seq., who says that the clauses do not eliminate the risk of error as a result not only of their appearance in places barely visible label but also the limited reading of labels by [the general public][consumers].

[188] See J. Massaguer Fuentes, Commentary on the Law of Unfair Competition, 1st ed., Madrid, 1999, p. 373 and 374.

[189] Interestingly in this regard, Community case law, the Judgment of the Court (First Chamber) of 19 April 2007 in Case C-381/05, which pitted Emmanuel De Landsheer SA against Interprofessionnel du Vin Committee Champagne and Veuve Clicquot Ponsardin SA. Also, most recently, the judgment of the same court of 14 July 2011 (C-4/10 and C-27/10, joined) in Case Cognac.

6.4 Evocation as unlawful autonomous conduct

6.4.1 On the concept of 'evocation'

As we have already explained, evocation is one of the behaviours prohibited by EU legislation in sub-paragraph (b) of Article 13(1) of Regulation (EC) No. 1151/2012. This is a behaviour that, in our view, is apt to cause in the market any of the three negative effects that have been mentioned. However, even if there is consensus on this point, the fact is that the absence of a normative notion or an authentic interpretation of that concept causes the added difficulty of answering the question: what is this behaviour? Or, to put in in another way, in which cases can we affirm that the evocation of a protected geographical designation has existed? This is, of course, a difficult legal issue, whose resolution demands the use of several interpretative tools.

In this context, we do not think that the location of the concept within the text of the Regulation is irrelevant in this hermeneutic work. On the contrary, its inclusion in this rule offers valuable information that helps to solve the question noted. Thus, its location next to two terms related to two unlawful behaviours (misuse and imitation) allows us to infer that EU legislation intends to make a kind of gradation of the infringement of a protected geographical designation and that the criteria used in the graduation have been precisely the closer or lesser approximation to the name of the protected geographical designation. Indeed, by including it in this section of the rule after mentioning 'any direct or indirect commercial use of a registered name' and referring to 'any misuse, imitation...' we could think – in good legal logic – that the drafter of the rule has laid down a progression of behaviours.

Applying this hermeneutic perspective, we believe that evocation would make reference to an unlawful autonomous behaviour different from the improper use – also referred to by Regulation 510/2006 as 'misuse' – and to imitation. Therefore, the improper use refers to that conduct consisting of the full reproduction of the name of the protected geographical designation (when used for the same kind of products which do not, however, possess the features contained in the specification), while imitation involves reproducing any of the terms of the name of the geographical designation which have most distinctive force in the market. Meanwhile evocation refers to a behaviour which consists of provoking consumer expectations on the relationship between a specific business offer and a protected geographical designation, which may arise resulting from any instrument such as geographical names, drawings or mere presentation of the business offer. For these purposes we can think, for example, of the importance of using indirect indications of provenance.[190] Indeed, we can consider that there is evocation when a link between a product and a protected geographical designation (to which the product does not belong) is induced in the recipients of a business offer, with a minimal intellectual effort and this irrespective of the elements which integrate such a connection, presentation of the business offer or the distinctive sign used. In these cases, even if the structure of the geographical designation is not reproduced, the reality is that this conduct induces in the consumers, the generation of associations that lead back to that in any case. Therefore far from making a direct reference to the geographical designation it is a different structure, which can be composed of elements of a different nature, which induces consumers to connect to the protected geographical designation.[191]

There is clearly a marked difference between these three behaviours included under letter(b) of the first paragraph of Article 13 of Regulation 1151/2012. Namely, the behaviours called 'misuse' and 'imitation' require a specific activity on the part of the infringer consisting in the total reproduction of the name or the part of it with a higher distinctiveness, while the behaviour comprising evocation does not revolve around the activities carried out by the offender but on its result on the market (generating a connection therewith by any means).

Moreover, this interpretative proposal of the rule is consistent not only with the first sub-paragraph (a) of the rule, which expressly refers to 'any indirect or direct commercial use of a registered name' but also with the definition of the verb 'to evoke' included in the dictionary of the RAE (Spanish Language Royal Academy) as 'to bring something to the memory or imagination' and in the online Oxford English Dictionary 'to call (a feeling, faculty, manifestation, etc.) into being or activity'[192]. Consequently, and from the consumers' perspective, it is evident that the evocation requires an intellectual effort which is not required with the concept of abuse or imitation. This is the opinion of, among others, Arean Lalín, who, when referring to the distinction between descriptive brand and suggestive or evocative brand, says that 'while descriptive designations communicate directly to the public the qualities or characteristics of the product, the suggestive designation forces the consumer to use his imagination and his reason to connect the brand with the product'[193]

---

[190] See on indirect geographical indications, C. Fernández-Novoa, International protection ..., cit., p. 2. More recently have dealt with them, M. Botana Agra, appellations of origin, cit., p. 19; M.M. Gomez Lozano, Appellations ..., cit, p.. 25 et seq. G. Sena, Il diritto dei marchi. Marchio Nazionale and Community marchio, 5th ed., Milan, 2007, p. 93, who refers to the geographic information derived from the use of a word in a foreign language.

[191] See C. Fernández-Novoa, Treaty ..., cit., p. 293 et seq. M. Botana Agra., 'The conceptual evocation of a word mark and a likelihood of confusion', ADI, t. VII, 1981, p. 249 et seq.

[192] www.oed.com; evoke (verb), definition 2(b).

[193] M. Lalín Arean, 'The ability of a brand name to become' ADI, t. V, 1978, pp. 479 et seq.

6.4.2 Evocation and semantic level protection of geographical designations

In view of the proposed interpretation, we are convinced that evocation is close to the semantic or conceptual similarity used in the field of trade marks. Indeed, alongside graphics and phonetic similarity verifiable between two distinctive signs, it is possible – as supported by doctrine and jurisprudence – that there is also a similarity at a conceptual level. For this reason, it is essential that the distinctive sign presents a particular meaning in common language, which is furthermore a manifestation of the real premise of the existence of a similarity in the conceptual level. In other words, this kind of comparison between two competing distinctive signs can only be sustained when the reference sign has a meaning in common language, regardless of their relationship with the specific product or service. This is, masterfully, defended by Prof. Fernandez-Novoa, when he says that:

> It is not possible to compare, from a conceptual point of view, all denominative marks. A conceptual comparison can certainly be possible in the hypothesis of marks consisting in suggestive or arbitrary designations. However, the comparison of designations in the conceptual level is improper in the hypothesis of marks consisting of capricious and fanciful designations.[194]

In this state of the matter, since protected geographical names have a specific meaning in common language because they refer to a specific territorial location, there should be no difficulty in accepting the possibility of a conceptual similarity with them, which will occur when the infringing sign or, where appropriate, the presentation of the particular business offer alludes to the same concept, either through the use of a specific place name or through an indirect indication of provenance.[195]

We can illustrate our interpretation referring to some cases adjudicated by Spanish and EU courts. Under EU Law, it is interesting to analyse the conflict between the PDO *'Parmigiano Reggiano'* and use of the term *'Parmesan'* to differentiate a particular type of cheese, a case which has been resolved by the judgment of the Court of the European Communities of 26 February 2008 (C-132/05), whose Paragraphs 47 and 48 affirm that:

> In addition, regardless whether the name 'Parmesan' is or is not an exact translation of the PDO 'Parmigiano Reggiano' or of the term 'Parmigiano', the conceptual proximity between those two terms emanating from different

[194] C. Fernandez-Novoa, Treaty ..., cit., p. 293.

[195] Interesting is for this purpose the judgment of the Provincial Court of La Rioja, to September 7, 2012, in which the existence of the violation of a designation of origin (Asparagus from Navarra) shown to induce the public to confusion about the product origin.

> languages, which was revealed in discussions before the Court, must also be taken into account.
> That proximity and the phonetic ... similarities referred to in Paragraph 46 above are such as to bring to the mind of the consumer the cheese protected by the PDO 'Parmigiano Reggiano', when he is confronted by a hard cheese, grated or intended to be grated, bearing the name 'Parmesan'.

Equally interesting is the judicial doctrine expressed by the Court of Justice in consideration of the Denomination of Origin cheese 'Feta',[196] as well as that relating to the Committee Interprofessionnel du Vin de Champagne, Veuve Clicquot Ponsardin SA.[197]

In Spain, we consider the case of the conflict between the Denomination of Origin 'Jamón de Huelva' and the commercial 'Origen Jabugo' that has been resolved by the Second[198] and Third[199] Sections of the Provincial Court of Huelva in the rulings of 8 and 20 February 2006, respectively. These are two judicial decisions which, in general terms, express our thesis because these judgments recognise that the toponym 'Jabugo', used to differentiate pork meat products, implies a violation of a Designation of Origin made up of the name of a province (Huelva, the province where the town Jabugo is located) and registered exclusively for one pork meat product, ham (*jamón*) from Huelva. In fact, the court considered that:

> the litis is born primarily from the disagreement of the plaintiff with the way in which the respondent uses the expression 'Origen Jabugo', in the legal name of the respondent company, as a means to identify in the market certain products derived from pork (hams, shoulders and others).[200]

The first of the Judgments concludes that:

> the latent conflict between the parties is clear, since they are both trading with the same genre of products and since the designation of origin administered by the plaintiff is broader and including the area mentioned in the controversial name, the competitiveness between the one and the other seal of quality

[196] Judgment of the Court of Justice (Grand Chamber) of 25 October 2005 (joined cases C-465/02 and C-466/02), the name 'feta'. It is exposed that 'the evidence presented to this Court also show that, in the Member States other than Greece, feta is commonly marketed with labels that evoke cultural traditions and civilization Greek.' It is legitimate to infer from this that consumers in those Member States perceive feta as a cheese associated with the Hellenic Republic, even if in reality it has been produced in another Member State.
[197] Judgment of the Court (First Chamber) of 19 April 2007 (Case C-381/05), which recognises that makes up a course of unfair advertising the evocation of an appellation of origin that the product actually does not possess.
[198] AC 2006/1104.
[199] JUR 2006/95250.
[200] The first legal basis of the judgment of the Provincial audience of Huelva on 8 February 2006.

> is obvious; and it follows that the sued company approached boundaries that were unclear when making use of it as a sign of identification of their products in a corporate name which was – at least to a greater extent than other possible names – at the limits of acceptability when it is used as equivalent to a brand.[201]

On the other hand, the Judgment of the Third Section affirms that there is a connection that is liable to create a risk of confusion between the two signs in conflict:

> since in the very name of the company the word, 'Origen' already appears and, together with the title of the town of Jabugo, in the province of Huelva, leads to a full confusion to consumers between both denominations, a company name and a designation of origin. This assertion is based on the social reality in which we live daily (...) as it induced the consumer in general to an obvious error between the activity of that company and a D.O., since as we have exposed a D.O. is a special kind of geographical indication, which applies to products that require a specific quality derived exclusively or essentially from the geographical environment in which they are produced. And the general public, consumers, understand, consider the D.O. as parameters expressive of the origin and the quality of certain products and these D.O.s require adequate legal protection by preventing the use as company names of geographical indications, which causes or provokes such a confusion.[202]

This interpretation was confirmed by the Civil Chamber of the Spanish Supreme Court in the order of 13 January 2009 refusing to admit the double appeal made by the company *Origen de Jabugo, S.L.* against the Judgment of 8 February 2006 of the Second Section of the Provincial Audience of Huelva. The fourth legal basis of this order contains the following statement in this regard:

> we meet an artificial ground for appeal and, therefore, non-existent, since the jurisprudence of this Court establishes the protection, in other States and for those nationals of States that are members of the Paris Union, of the trade name without deposit or registration, equating them to the nationals of

[201] Second legal basis of the decision referred to in the previous footnote.

[202] Third legal basis of the judgment of the *Audiencia Provincial* (High Court) *of* Huelva of February 20, 2006. However, in the judgment of the *Audiencia Provincial* (High Court) of Vizcaya, Section 4ª, on April 19, 2011 (third and sixth legal bases),even if there were different legal premises, the conclusion was diametrically opposite in relation to an alleged conflict with the denomination of origin, Idiazabal. The decision has to be different as 'the words that consist of the names of 'Etxegarai' and before 'Etxegarate', cheeses that are in the area of production of Idiazabal cheese do not match the name of any of the considered places of special protection by the regulations of the DO use prohibited (...)' 'and that the geographic scope of the production area and the names of the municipalities comprising are barely known to consumers, so it is unlikely that the name controlled cheeses evoque in the consumer half the products covered with the DO Idiazabal'.

> these States, as per Article 8 of the Convention of the Union of Paris, provided they fulfil the conditions and formalities as required of nationals, conditions that are understood by the Audience that are not met in this specific case to the extent where the use of the name of the now appellant is considered an act of unfair competition since it leads to confusion to consumers regarding the quality of the product, because it evokes a denomination of non-existent origin, therefore it would incur a relative prohibition that would determine that the legal requirements were not fulfilled and, consequently, the intended protection could not be delivered.

6.4.3 Unfortunate initial exegesis by the CJEU and its recent correction at the request of the OHIM

This conclusion - consistent with the breadth of the drafting of the principle contrasts with the interpretative line advocated by the Court of Justice of the European Union. In effect, having to face the concept of 'evocation' in order to adjudicate on different cases submitted to the Court, the Court has sustained a restrictive exegesis of the same, by proposing an interpretation which considerably reduces the scope of application of the precept, resulting in a correlative restriction in the protection of qualified geographical designations.

This interpretative line begins at the end of the 1990s, when the Court was confronted with the task of interpreting the aforementioned concept. Specifically, the first decision rendered by the Court in this regard is contained in the Judgment of 4 March 1999 (C-87/97), pronounced in the case Gorgonzola/Cambozola, where there is a double assertion with - in our opinion - uneven success. Indeed, when faced with the interpretation of Article 13(1)(b), the Court affirms - Paragraphs 25 and 26 - that:

> Evocation... covers a situation where the term used to designate a product incorporates part of a protected designation, so that when the consumer is confronted with the name of the product, the image triggered in his mind is that of the product whose designation is protected.
> 26. As the Advocate General states in points 37 and 38 of his Opinion, it is possible, contrary to the view taken by the defendants, for a protected designation to be evoked where there is no likelihood of confusion between the products concerned and even where no Community protection extends to the parts of that designation which are echoed in the term or terms at issue.

More recently, the Court of Justice has dealt with this matter in the case of Parmigiano Reggiano/Parmesan, which we have mentioned before, where, although the Court alludes to the evocation as a form of 'conceptual proximity', the interpretative line initiated previously is ratified. In particular, Paragraphs

44 and 45 reproduce *the* double passage of the first pronouncement exactly as it was transcribed above.

However, the content of these considerations integrated into both judgments does not induce us to the same opinion. Indeed, while we consider positively the separation carried out by the Court between the evocation of a geographical name and the aggravation of a likelihood of confusion, since one can exist without the other, and therefore to be evocation if it triggers only a misappropriation of reputation, we believe the interpretation made of the concept of 'evocation' is unfortunate. Because, if we make a careful reading of the above-mentioned merits of the case, it can be inferred that this concept relies on the cumulative presence of two requirements: in addition to the associative effect between the two signs in question made by the average, normally informed, consumer, reasonably attentive and circumspect, an adequate cause is required in order to generate that dire consequence, namely, a structure of the offending sign that incorporates part of a geographical name.

This second requirement, by limiting the semantic cover of the concept of evocation, reduces the protection of recognised geographic designations. We must recall in this respect that – as we stand for and following the interpretative line of the Spanish Supreme Court – evocation requires only the provocation of a connection with the protected geographical designation regardless of the elements of its structure. Therefore, there is no doubt that if we follow the exegesis proposed by the European Court of Justice, the protection in the EU of geographical designations is not working when facing an injurious behaviour that is very common in practice, the adoption of a toponym integrated within the territorial demarcation of a protected geographical designation as the distinctive sign that will differentiate products identical or similar to the protected product.

For this reason, the first decision taken by the Office for the harmonisation of the internal market in defence of a protected geographical name against the attempted registration of a Community trade mark, should be assessed positively. And we applaud it because, giving a response to the situation mentioned above, it allows a first step to redirect the unfortunate interpretation of the concept of evocation proposed by the Court of Justice. We refer to the decision taken by the Division of Opposition of the OHIM on 27 January 2011, in the case where the regulatory Council of the protected geographical indication 'Valencian Citrus' objected the applicant for the trade mark 'Castellón Clementines'. In this decision, the registration of the trade mark was refused on the grounds that:

> the contested Mark uses two names protected by the aforementioned 'Valencian citrus' PGI: 'Castellón' and 'Clementines', because with these names we are referring, on the one hand to the geographical name of one of the provinces of Valencia and, on the other hand, to the citric fruit protected:

Clementina, called Clementine in French. Furthermore, the graphic part also reproduces the figure of a citrus. On the other hand, Article 13(1)(b) CTMR applies even when the signs in question are not confused, the mere evocation of the indication of origin is sufficient. Thus a sign reminiscent of another when a person manages to bring it to memory. Therefore, the contested mark is reminiscent of the PGI 'Valencian Citrus' or at least the consumer immediately relates it to such products. Therefore, an association on the part of the public occurs so that, when contemplating the product, it is thought of as an image referring to products designated by the protected indication of origin. So based on this evocation the applicant can misuse the reputation that the 'Valencian Citrus' PGI enjoys.

7. The working of the food alert system in Germany

Market-related measures, RASFF, publication of information

Nicola Conte-Salinas

7.1 Overview

This chapter covers possible consequences of a violation of food law or complaints (*Beanstandung*) by authorities. Besides administrative and criminal sanctions, the main consequences are a) market-related measures, and b) the publication of information by various means.

The chapter begins with an overview concerning the German legal situation and jurisdiction concerning market-related measures, i.e. recalls and withdrawals. With respect to the publication of information, an overview over the functioning of the RASFF system in Germany is given, as well as a quick overview over particular German law.

7.2 Market-related measures

Recalls and withdrawals concern food that has left the immediate control of the initial food business operator, and are therefore dealt with under the headline of market-related measures.

7.2.1 Section 39 LFGB

Section 39 of the German Food and Feed Code (*Lebensmittel und Futtermittel Gesetzbuch* – LFGB) lays down responsibilities and measures to be taken by competent German authorities. According to Section 39 Paragraph 2 sentence 2 No. 4 LFGB, for example, authorities can initiate market-related measures (recall, withdrawal), if necessary to confirm or dispel a probable cause of a violation, to remove a proven non-compliance or to prevent future non-compliance, and to protect from health dangers or from misleading information.

This means, the right of authorities to demand a recall/withdrawal, goes beyond the duties of food business operators according to Regulation (EC) No 178/2002: German authorities are in principle entitled to order product withdrawals or product recalls in cases of food that is considered to be injurious to health,[203]

[203] According to Article 14(2)(a) Regulation (EC) No. 178/2002.

unfit for human consumption,[204] but also if there is no imminent health risk for consumers, for example in cases of wrongful labelling.

7.2.2 Case study

The Higher Administrative Court (OVG) of Northrhine-Westphalia explicitly decided in August 2008,[205] that on the basis of Section 39 Paragraph 2 sentence 1 in connection with sentence 2 No 4 LFGB, authorities can order a withdrawal or a recall of a product from the market, even if the non-compliance in question is not health-related, but concerns incorrect labelling of the product.

In May 2009, the Higher Administrative Court (VGH) of Bavaria also confirmed,[206] that the existence of a health risk is not necessary in order to legally order a product to be withdrawn. The case concerned food contact materials that led to a change in the organoleptic properties of the food.

The Higher Administrative Court (VGH) of Baden-Württemberg had to decide a case of illegal nicotine residues in a spice mix in March 2010. It was suspected that the eggs that were used in the spice mix contained nicotine, as nicotine was illegally used to disinfect life animals in the chicken farms concerned. A health risk for consumers was ruled out. The company producing the spice products immediately stopped using the powdered eggs and stopped the delivery of finished spice products containing the egg powder in question. However, a certain amount was already used in its products and sold to its customers. In the case at hand, the Court left the question unanswered, whether authorities can legally order a food that is contaminated with residues of unauthorized pesticides to be taken off the market, even if no health risks are present. But due to the circumstances of the case at hand, the order of recall/withdrawal was considered disproportionate.

7.3 Publication of information

Today, there are various legal grounds according to which information can be published (Figure 1). As regards food and feed, the RASFF system operates throughout the EU, and is undoubtedly an important instrument to transmit information. But also national laws and systems concerning the publication of information can have an enormous effect on the food businesses concerned. Among these national measures and systems is the Consumer Information Act (*Verbraucherinformationsgesetz* – VIG), State Press Acts, Section 40 LFGB, as well as information on hygiene compliance ('traffic lights', 'smiley systems', etc).

[204] According to Article 14(2)(b) Regulation (EC) No. 178/2002.
[205] File No.: 13 B 1022/08.
[206] File No.: 9 CS 08.3300.

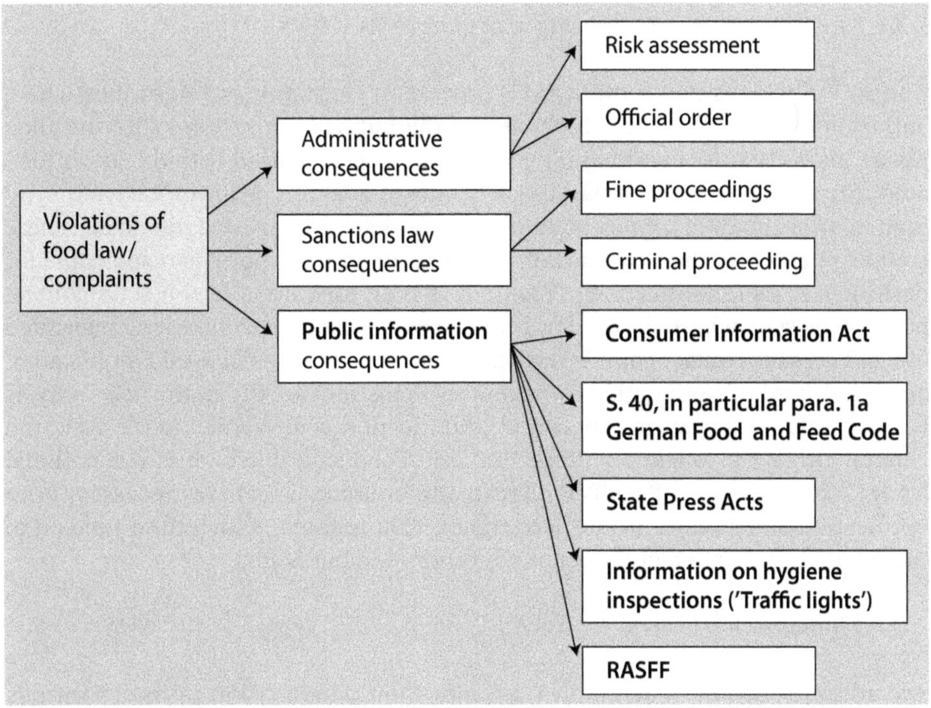

Figure 1. Various legal grounds to publish information

The following subsections explain how the RASFF system is implemented in Germany, focusing on the question whether legal remedies against the publication of information in the RASFF system are available. This section concludes with a short overview over national measures according to which food information can be published.

7.3.1 RASFF

Articles 50 – 52 Regulation (EC) No. 178/2002 are the EU basis for the RASFF system. Implementing measures have been laid down in Regulation (EC) No. 16/2011.

Germany consists of 16 federal states (*Bundesländer*). The competence for the official food surveillance lies with these federal states. The General Administrative Rule Rapid Alert System (AVV SWS – *Allgemeine Verwaltungsvorschrift Schnellwarnsystem*) was enacted in order to ensure uniform application of these rules. The AVV is binding for the administration only (internal administrative rule), but not binding on the courts.

7.3.1.1 RASFF process in Germany according to AVV SWS

The AVV SWS lays down the RASFF process in Germany and describes which authorities are involved in which step of the publication process (the so-called upstream procedure; Figure 2). The process is usually started with the lower food authority that detects a food-based risk to human health. The lower food authority then informs the state contact point of the state where the risk was detected. These contact points may differ from state to state. This contact point is responsible for drafting a RASFF notification. This draft is then forwarded to the state contact point of the state where the food business operator has its headquarters/registered office. This state contact point is responsible for the decision if a RASFF notification and what kind of notification is necessary. This means, the main risk analysis is carried out at this stage. The RASFF notification is forwarded to the BVL, the Federal Office of Consumer Protection and Food safety, which is the national RASFF contact point. The BVL checks and completes (where necessary) the notification and communicates it to the EU Commission, which then sends it to the national state contact points of the other Member States.

7.3.1.2 Initialisation criteria for a RASFF notification, Section 7 AVV SWS

According to Section 7 Paragraph 1, a notification is required in cases of a serious direct or indirect risk for human health. Paragraph 2 lists cases in which a direct or indirect risk for human health is irrefutably assumed. This means, authorities have no discretionary power, if these cases are established. Circumstances, where authorities explicitly have to assess whether a direct or indirect risk for human health exists, are listed in Paragraph 3.

RASFF process in Germany according to AVV SWS:

1. Lower food authority
   - Detects food based risk to human health
2. State contact point of the state where risk was detected
   - Drafts RASFF notification
3. State contact point of the state where food business operator has its headquarters/registered office
   - Decides if a and what kind of RASFF notification is necessary (main risk analysis)
4. BLV (Federal Office of Consumer Protection and Food safety) = national RASFF contact point
   - Checks and completes (where necessary)
5. EU Commission

Figure 2. Upstream procedure of RASFF in Germany.

An irrefutable assumption of a serious direct or indirect risk for human health exists, inter alia:

- If a food contains a substance not allowed in food (according to EU or German law).
- If a food exceeds ADI (Acceptable Daily Intake)-values of pesticide residues.
- If foods are/contain unauthorised GMOs.
- If foods are unauthorised Novel Foods.

However, these criteria do not necessarily correspond with an actual health risk. This means that according to German administrative practice, RASFF notifications may be issued without an actual health risk.

7.3.1.3 Legal remedies

Food business operators are interested in preventing the transmission and ultimately the publication of RASFF notifications, since more often than not, these notifications result in financial loss as well as image losses. These might affect whole industries, as was shown during the EHEC outbreak in 2011 ('Spanish cucumbers') or individual food businesses. Although the RASFF notification itself is anonymous, it can have substantial external effect, since in connection with such a notification, authorities carry out investigations along the supply chain, including customers. Therefore, it is of utmost importance for food business operators to determine whether it is possible to prevent the transmission of an alert by legal means.

According to literature and case law, the draft by the state contact point of the state where the risk was detected *cannot* be prevented, as the draft has no legal effect. The transmission of the notification by the BVL to the EU Commission cannot be prevented (if the notification is complete and consistent), as the BVL only verifies the completeness and consistency of a notification. The role of the BVL is similar to the role of the EU Commission in the RASFF system.[207] The question, whether the decision to issue a RASFF notification by the state contact point of the state where the food business operator has its headquarters/registered office, is illustrated by three German cases (application for interim measures).

The first two of these cases concern the RASFF alert notification due to the presence of listeria monocytogenes in premade fried chicken schnitzel.

The first Court who had to deal with this case was the Bavarian High Administrative Court (VGH München). The Court decided that a product warning can only be carried out, if it is necessary, since such a warning can potentially infringe basic rights of food business operators. However, in the case at hand, it deemed such

[207] See Zechmeister/Van der Schoot, ZLR 2008, 583; VGH München, M 18 K 11.1445; cf. Decision of the Court of First Instance Case T-177/02'Malagutti Verzinhet/Commission.'

information necessary due to the imminent risk to human health (despite the fact that the product was already past its best-before-date and note on the labelling that the product needed to be heated before being consumed).

The High Administrative Court of Lower Saxony (OVG Niedersachsen) confirmed the finding of the VGH München, although it partly differs in its reasoning: Since the RASFF notification is anonymous and since the notification itself does not have any (external) effect in the other Member States, but requires actions from the relevant public authorities, the Court doubted whether the action was admissible. The Court did not decide this question, since it considered the claim unfounded, as the Bavarian High Administrative Court had done before.

The latest decision that dealt with the admissibility of legal remedies against RASFF notifications concerned a tea of stevia leaves that was used as an ingredient of a yoghurt. The competent authorities qualified tea of stevia leaves as a novel food and intended to initiate a notification. With an application for interim measures, the food business operator concerned tried to prevent such a notification. The Bavarian High Administrative Court (VGH München) granted interim relief, based on the following reasoning:

If unjustified, a RASFF notification can infringe basic rights of food business operators, due to possible substantive economic and image loss. Further, the Court found the claim to be well founded, as it considered yoghurts containing tea from leaves of *Stevia rebaudiana* Bertoni not to be Novel Food.

To sum up, there is no uniform case-law regarding the prevention of a rapid alert notification. It is controversial, if an infringement of rights is at all possible in these cases. Admittedly, the RASFF notification as such is anonymous, but it also has to be taken into account, that in connection with rapid alerts, intensive investigations by public authorities take place along the supply chain. Customers tend to take precautionary measures and completely stop the marketing of the food concerned. Besides this financial loss, these measures can result in a substantial image loss of the food business operator concerned. Further, the possibility exists to use laws that grant a right to information to re-personalize the information in the RASFF system.

7.3.2 National laws

7.3.2.1 Consumer Information Act – VIG

The German Consumer Information Act (VIG) allows for extensive possibilities to grant access to information concerning 'inadmissible deviations from food law' (any form of subjective/personal fault is deemed unnecessary). These rules, however, are outside the scope of this chapter.[208]

7.3.2.2 Section 40 LFGB: publication of information

Section 40 of the German Food and Feed Code (LFGB) provides for various rights and duties to inform the public.

For example, Section 40 Paragraph 1 LFGB lays down the right to inform the public not only where food is injurious to health, but also if food is unfit for human consumption, in particular because it is nauseating, is or has been distributed in significant quantities or where, because of its specificity, it has been distributed only in small quantities but over a relatively lengthy period of time. Section 40 LFGB therefore goes beyond the cases covered by Article 10 Regulation (EC) No 178/2002. The Regional Court (LG) of Munich asked the Court of Justice of the EU for a preliminary ruling in December 2011 (*Berger Wild* – C-636/11), in order to determine whether this law violates Article 10 of Regulation (EC) No 178/2002. In its ruling, the CJEU decided that Article 10 of Regulation (EC) No 178/2002 must be interpreted as *not precluding* national legislation allowing information to be issued to the public mentioning the name of a food and the name or trade name of the food manufacturer, processor or distributor, in a case where that food, though not injurious to health, is unfit for human consumption. The CJEU ruled, that the national authorities may inform the public in these cases in accordance with the requirements of Article 7 of Regulation No 882/2004.

Warnings according to Section 40 LFGB are also published on the internet.[209]

7.4 Conclusion

These findings give rise to the question whether formal judicial procedures are gradually being substituted by pressure from media. Is this the beginning of 'enforcement by information'?

[208] For more information, see *Grube/Immel/Wallau*, Verbraucherinformationsrecht, 2013; Theis, DVBl 2013, 627 (on the VIG and Section 40 para 1 a LFGB).
[209] See: www.lebensmittelwarnung.de.

8. Legal problems arising out of the food alert system in Spain and Europe

Vicente Rodríguez Fuentes

8.1 Introduction

Food alerts are designed as a tool for the communication of food risks. Food alerts are instant, rapid, like an alarm bell intended to be raised in case of serious risk to public health. The system is powerful and its internal working has been carefully designed by a specific Regulation[210] that determines who transmits and receives the information and how it is to be transmitted.

However, no similar attention has been paid to describe when the alert is necessary or, more exactly, when the consequences of the alert are proportionate, *necessary* in the words of the European Court of Justice[211]. As a matter of fact, the regulation of food alerts presumes that its scientific justification and economic impact are correct.

A food alert is more than an internal network of communication; it is an administrative action whose effects can potentially be disastrous for affected companies. The impact of the food alert can very easily surpass that of any possible sanction. We only need to remember the alert on *Escherichia coli* and Spanish cucumbers which is said to have led to losses of 20 million euros per day (affecting not only cucumbers and not only Spanish products). Many companies affected by a food alerts went bankrupt and disappeared.

But in spite of this potential impact, no specific procedure is required to adopt the decision to communicate the risk. The recall of products is not considered a sanction nor an expropriation. The food alert is issued without prior contradiction, no presumption of innocence applies, no typical procedure divided into different stages where parties can defend themselves, present their arguments, use the pieces of evidence that they have at their disposal, exist and so on.

Does due process applies to the food alert system?

The fact is that, due the legal nature of a food alert, its economic consequences, its relevance to public opinion (affecting national culture, as is food), companies affected by a food alert must defend their rights by means of legal instruments that are not always adequate to this new reality and that pose a number of specific

[210] Regulation 16/2011 in relation with Articles 10 and 50 of Regulation 178/2002.
[211] ECJ C 217/99.

legal problems that I would like, very briefly, to outline, without intending to give any answers (although of course, I have my opinions).

8.2 Some of the legal problems of food alerts

8.2.1 The problem of legal standing

The first problem that, as a lawyer, I would face would be that of legal standing to contest a food alert, since without legal standing no legal/judicial debate on the alert is possible.

We all know that to contest an administrative activity, which a food alert is, all legal systems require contenders to have a standing in the case. The actions of the European Commission (like transmitting a food alert through the rapid alert system) can be only challenged by the persons to whom this decision is directly addressed. In Spain, you are required to demonstrate that you have a legitimate interest in the case and, I guess, most if not all legal systems require a similar interest in the case.

But food alerts can be confidential, sometimes no company or brand is mentioned and the effect of food alerts can easily be felt by companies who do not deal with the product mentioned by the alert, just because they happen to manufacture a similar product or belong to the same country and are affected by the panic some food alerts unleash in the market. On top of this, if the food alert is international, the content of the alert and its effects are often the result of the original alert plus the particular interpretation of the different national authorities.

The result is that, in some cases, affected companies are not the direct subject of the text of the alert and have first to demonstrate how the alert has affected them directly and demonstrate a causal link between the alert and its effects, in a context where these effects are amplified by the reaction of the market. This is a legal task that is not always simple, because a very strict interpretation of who is affected by the alert could limit the right of some parties to legally act against it.

8.2.2 The problem of identifying the responsible authority

A food alert can involve three different levels of public authority. I will call them the initiating, the transmitting and the implementing authorities. These authorities are numerous, based in many different countries and connected through the network of the rapid alert system.

Therefore, one of the main legal problems that face a company affected by a food alert is to identify the contender; who is responsible for the consequences of the alert. In my experience (particularly when a food alert appears disproportionate), different public authorities tend to pass the responsibility onto each other, by saying – the initiating authority – that they simply communicated to the network an information that seemed to justify the alert, but cannot be held responsible for the reaction in other countries or markets. Then, the transmitting authority justifies its action by saying that it has to act when a serious health risk is communicated to them, without questioning the information due to its urgency. In the same manner, implementing authorities affirm that they have no choice but to react and implement the alert that has been communicated to them.

Therefore, if we want to contest the alert (because we believe that is not justified), if we want to stop it and reverse its effect, who do we challenge? Can we afford to sue a foreign authority or to start several legal proceedings in different national jurisdictions? And what would happen in case of contradictory Court decisions? The answer to these problems in the context of a national legal system is relatively straightforward, because there are rules to determine the appropriate jurisdiction taking preference over the rest. But this does not happen in a food alert, an international, complex and instant administrative action.

The careful – and economically affordable – choice of the responsible authority, in a case where the consequences of the alert could be attributed to the joint action of several authorities, is another difficult legal problem that we face when confronting a food alert.

8.2.3. The control of legality of the alert

The legality of the alert can be assessed, like with any other administrative decision, by its correct formal production and by its legal premises.

1. A food alert must be formally justified (in the sense that a sufficient documentary file should exist to help us understand and to question the reasons for the alert) and must be issued by the competent authority, which communicates the food alert through the food alert network. Moreover, it can be the case that the worst effects of a food alert are produced by its communication to the public because this damages the reputation of the product and can produce panic on the market with terrible economic consequences.
   Is a press conference an administrative action that can be challenged at Court? How do we control that this communication is made by the competent authority? Is any authority more competent or are all equally competent? Has the issuing authority a greater interest and responsibility than the rest?
   As a matter of fact, it is frequently the case that the authorities tend to communicate more when the food alert is provoked by a foreign product. In my opinion this is not only for economic reasons – to be more careful with local interests – but also because a food alert is a powerful means of creating public debate, a debate where both the terms of the debate and the argument are favourable to the issuing authority and were the stereotypes concerning

foreign food always play a role (a foreign food producer easily becomes a foreign food product and a foreign food product easily becomes all foreign food products coming from that country or, at least, those food products in that country loosely related to the food affected by the alert).

Again, we face an important legal problem. How do we defend reputation when it is affected by information unleashed by a food alert? How do we control the formal legality of that information?

2. A second question is: can we challenge the reasons given for the alert?

   A food alert is only possible in response to a serious risk to health. Any other kind of risks – that is, less serious – would justify a different protective action but not a food alert. In a similar way, the extension and scope of the alert, the number of products and countries affected, the content and extension of its communication to public opinion (if there is a need to make the alert public at all) is only justified in case of a serious health risk that cannot be dealt with by less drastic action. This is a requirement of the principle of proportionality, a principle that must be interpreted also in accordance with the precautionary principle, which permits to take action even if the risk is not fully demonstrated, but does not cover actions based on a merely hypothetical assessment of risks. To revise the factual grounds of an administrative action in the field of food risk and public health, where important personal rights are at stake, where authorities enjoy a great degree of discretion and have – justifiably – great powers of intervention, is a legal problem formidable in itself. But at least in this case, this is a legal debate that administrative and judicial procedures are ready to deal with and leave the affected company the possibility to argue and challenge the correctness of these grounds.

8.2.4 The problem of halting the alert

In any case, it is clear that initiating, implementing or refusing to withdraw a food alert is an administrative decision that can be challenged at the appropriate Court and there are judicial precedents that support this position. As far as I know, a food alert has been annulled in Spain by the Spanish Supreme Court[212] and, somehow similarly, the decision of the Commission not to withdraw an alert was provisionally revoked by the European Court of Justice[213].

However it does not seem legally feasible to obtain an injunction that orders all administrations in all different countries to stop the alert. In fact, Regulation 16/2011 expressly rules that a food alert can be modified or withdrawn, but always with the consent of the initiating authority.

[212] STS of 27/06/2007 ASOLIVA (VRF was the legal counsel in this case)

[213] Order of 12 September 2006 in Case T-212/06 R *Bowland Dairy Products* v *Commission*.

Moreover, another legal problem we face when confronting an alert is that of its legal nature. Since the alert is a mere internal communication through a rapid communication system, it could be considered an interim rather than a final act. Sometimes the authority has denied that it is an administrative act at all. This interpretation on the nature of the food alert has been defended by the administration in some cases where I have intervened. In their view, a food alert is something more than a mere procedural step, but is not a final administrative decision but a *tertium genus,* an intermediate act that cannot be challenged in the administrative courts. In this interpretation, the Courts could only review the administrative implementation of the alert, for example, a recall of products. This interpretation (that would leave out of the world of the law, so to speak, the consequences of the alert provoked by the reaction of the market, or the decision to call for a press conference informing of the alert) was rejected by the Spanish Supreme Court[214].

But it is a reality that the loose legal nature of the food alert makes challenging it more difficult than other administrative decisions, whose consequences are less aggressive than those of the food alert.

8.2.5 The problem of removing the effects of the alert

I have no doubt that initiating, implementing or refusing to withdraw a food alert are administrative decisions that can be successfully challenged in the appropriate Court, as has been the case at the Spanish Supreme Court and the ECJ.

However when it comes to removing the effects of the food alert, compensating the damages caused to the affected company, or establishing the extra-contractual liability of the State (or the European Institutions) for a food alert, the answer in the Courts appears to be different. So far, to my knowledge, there have been few cases in Europe where the Court has granted compensation to affected companies.

The European Court of Justice has up to this moment refused to compensate the damages caused by an unjustified food alert, in the Bowland[215] dairy and Malagutti[216] cases. Similarly, the Spanish Supreme Court annulled an alert but refused to compensate for damages on virtually – in my opinion – opposite merits[217], in the pomace olive oil cases. Something similar happened when the Spanish Supreme Court annulled a communication of risk affecting meat products but refused to compensate damages alleging the lack of a causal link[218].

[214] STS of 27/06/2007.

[215] ECJ, Judgment of the Court of First Instance (Fifth Chamber) of 29 October 2009, case T-212/06.

[216] ECI, Judgment of the Court of First Instance (Second Chamber) of 10 March 2004, case T-177/02.

[217] STS 4/03/2009 and others, particularly the STS of 20/10/2009 which is the first one repealing a Judgment of the lower court granting damages for the consequences of the food alert.

[218] STS of 3/03/2009.

The fact is that, in my opinion, Courts tend to be reluctant to admit the extra-contractual liability of the State, in cases like food alert where the administration enjoys a great degree of discretional power in the protection of public health, a right that takes preference over economic considerations. It seems to me as if the Courts were, somehow, reluctant to limit the ability of the administration to react in cases of risk to public health by making the Administration responsible for an action that the Court has, otherwise, no problem declaring illegal or disproportionate. It seems as if it is one thing to decide that the administrative decision is not justified but a different matter to adjudicate on who is to bear the economic consequences of this unjustified administrative decision.

However, even if the right to health is – without any doubt – more important than the right to property, a fair balance of interests must always exist. Companies are responsible for the safety of the food products they place in market, but no one should be deprived of their property without sufficient reason and adequate and prompt compensation.

To make matters more complex, the damages caused by a food alert are very frequently (particularly in the worst cases) the consequence of the reaction of the market, sometimes amounting to panic, rather than the result of the direct intervention of the authorities. Therefore, another legal problem comes from the existence of a causal link between action and damages, a causal link which must be clearly demonstrated and which the European Court of Justice has, so far, interpreted in very strict and limited terms.

8.3 Conclusion

By way of conclusion, the Food Alert represents a new kind of administrative activity with very transcendental effects. This administrative activity has not, in my opinion, been properly regulated nor been remedied to its consequences (if unjustified) been properly taken into account by administrative law. The traditional legal and judicial means of control of governmental activities and eventual compensations are, in my opinion, inadequate to this new reality.

A legal regulation of food alerts where the procedure, justification and compensation were better defined would do no harm to public health protection and would favour legal certainty and justice. Within the rule of law, the ends do not always justify the means.

9. The case in the USA: safety alerts and the Reportable Food Registry

Ricardo Carvajal

In the United States (U.S.), the regulatory landscape for public notification of the existence of a potential food safety issue has changed significantly in recent years, particularly for foods other than meat and poultry. Such foods fall under the jurisdiction of the U.S. Food and Drug Administration (FDA). Historically, FDA learned of a potential food safety issue as the result of a consumer complaint or a report from a public health agency charged with conducting surveillance for foodborne illness. FDA could also learn of potential food safety issues based on voluntary reports from manufacturers or distributors. In 2007, the Federal Food, Drug, and Cosmetic Act (FDC Act) was amended to make reports from manufacturers or distributors mandatory under certain circumstances.[219]

The 2007 amendment to the FDC Act established a mandatory reporting system called the Reportable Food Registry (RFR).[220] The RFR is an electronic reporting system that is intended to provide a reliable mechanism for FDA to track patterns of adulteration in food, and to help FDA target its limited inspection resources to maximize public health protection. Failure to comply with reporting requirements under the RFR is a prohibited act that is punishable by civil and criminal penalties.[221]

The RFR requirements are triggered when a 'responsible party' determines that a food is a 'reportable food.' A 'responsible party' is defined as the person who submits the registration to FDA for the facility at which food was manufactured, processed, packed, or held.[222] Registration of facilities is required under Section 415 of the FDC Act, which applies both to domestic facilities and to foreign facilities that manufacture, process, pack, or hold food for export to the U.S.

A 'reportable food' is a food for which there is a reasonable probability that the use of, or exposure to, the food will cause serious adverse health consequences or death to humans or animals.[223] Once a responsible party determines that a food is reportable, that party must submit an RFR report to FDA as soon as practicable, but in no case later than 24 hours after determining that the food is reportable.[224] By implication, the obligation to determine whether a food is reportable falls on the responsible party. Therefore, when a responsible party receives information which

[219] Food and Drug Administration Amendments Act of 2007, Pub. L. 110-85, 121 Stat. 823, (Sept. 27, 2007).
[220] FDC Act § 417.
[221] Id. § 301(mm).
[222] Id. § 417(a)(1).
[223] Id. § 417(a)(2).
[224] Id. § 417(d)(1)(A).

suggests that a food might be reportable, the responsible party should promptly gather the information needed to determine whether the food actually meets the standard of reasonable probability of serious adverse health consequences or death. Also, if the adulteration may have originated with the responsible party, then that party must promptly investigate the matter to determine the cause of the adulteration.[225]

There is a limited exception to the RFR reporting requirement. There is no obligation to submit a report if all three of the following circumstances apply: (1) the adulteration originated with the responsible party; (2) the responsible party detected the adulteration prior to any transfer of the food to another person; and (3) the responsible party corrected the adulteration, or destroyed or caused the destruction of the food.[226]

Once submitted, an RFR report can't be withdrawn or expunged, and is subject to the Freedom of Information Act (FOIA) and FDA's regulations implementing that law. Under the FOIA, any person can submit a request to FDA for records associated with an RFR report. However, FDA is prohibited from disclosing to the public certain types of information in those records. For example, any information that constitutes a trade secret or confidential commercial information would not be subject to disclosure.[227] The FDC Act recognizes that submission of a report is not an admission of adulteration or of causing injury, and any report can include a disclaimer to that effect (e.g. 'our submission of this safety report is not an admission that the product is adulterated or that it caused or contributed to a death, serious injury, or serious illness').[228]

Upon receipt of an RFR report, FDA promptly reviews and evaluates the submission to determine whether the food in fact qualifies as a reportable food. Recent and typical examples of reportable foods include foods contaminated with Salmonella or other foodborne pathogens, and foods that contain undeclared major food allergens. Less typical examples include low-acid canned foods that contain Clostridium botulism toxin due to under-processing, and baby foods that pose a choking hazard.

If FDA determines that a food qualifies as a reportable food, then FDA coordinates with the responsible party to ensure that its suppliers and customers are notified so that they can cease distributing any potentially affected food and submit their own RFR reports. Typically, FDA also expects that the responsible party will conduct a voluntary recall that extends to the consumer level. Such recalls are usually accompanied by a notice to the general public in the form of a company press

[225] Id. § 417(d)(1)(B).
[226] Id. § 417(d)(2).
[227] 21 C.F.R. § 20.61.
[228] FDC Act § 417(l).

release. FDA ordinarily gives the company conducting the recall an opportunity to draft the press release announcing the recall, in consultation with FDA. The press release is expected to specify the nature of the risk and note if there have been injuries. Once FDA agrees with the content of the press release, then the press release can be issued and will be posted on FDA's website. If a company refuses to issue a public notice, then FDA can be expected to issue its own public alert. Also, if a company refuses to conduct a voluntary recall, then FDA can order the company to conduct a mandatory recall.[229]

In some instances, FDA and other public health agencies might choose to issue their own alert even when a responsible party has issued a public notice and instituted a voluntary recall. The FDC Act specifically authorizes FDA to issue an alert or notification with respect to a reportable food using information from the RFR as necessary to protect public health.[230] FDA is also authorized to disseminate information regarding food in situations involving imminent danger to health, or gross deception of the consumer.[231] In the case of a mandatory recall, FDA is authorized to require a press release, alert, and public notice of the recall.[232]

For FDA, the decision of whether and when to issue a public alert that implicates a specific food can be difficult. In some cases, the fact that there is imminent danger to health may be evident, but the specific food at fault might not be. FDA might be left to rely on epidemiological evidence, which can lead to erroneous attribution. For example, in response to an outbreak of salmonellosis in 2008, FDA initially announced that the outbreak appeared to be linked to certain types of tomatoes. Shortly thereafter, FDA issued a second announcement that cleared tomatoes grown in certain geographic areas. Eventually, FDA announced that tomatoes were not associated with the outbreak, and that peppers were the culprit. By then, the tomato industry had suffered significant losses.

The potential to recover damages from the U.S. government in cases of erroneous attribution is very limited. The doctrine of sovereign immunity generally precludes lawsuits against the government. However, the Federal Tort Claims Act (FTCA) creates exceptions to the doctrine of sovereign immunity in certain cases involving negligent acts.[233] A claim under the FTCA must first be filed as an administrative claim with the U.S. Department of Health and Human Services, FDA's parent agency.[234] If the administrative claim is denied, then the company can sue in federal court. Generally, such lawsuits are expensive to pursue and difficult to win, in part due to the burden of demonstrating negligence.

229 Id. § 423.
230 Id. § 417(c).
231 Id. § 705(b).
232 Id. § 423(g)(1).
233 28 U.S.C. §§ 1346(b), 2671-2680.
234 42 C.F.R. Pt. 6.

Partly because of cases of erroneous attribution and the very limited relief offered by the FTCA, there has been discussion of possible measures of compensating companies that are adversely affected by inaccurate government alerts. These measures include a government administered compensation program, preferential treatment in government contracting, situational legislation (e.g. proposed legislation that would have provided $100 million for tomato growers), insurance programs, and governmental promotional campaigns designed to counter prior adverse publicity. However, the current outlook for agreement on any of these measures is dim, and it is not clear that any of them would be available to foreign companies.

From a foreign company's perspective, an FDA Import Alert presents the ultimate challenge in the realm of safety alerts. If it appears from FDA's examination of a food product offered for import that the product was manufactured, processed, or packed under insanitary conditions, or that the product is otherwise adulterated or misbranded, then FDA can detain the product at the border and place the product on Import Alert.[235] That action notifies FDA field staff that all shipments should be detained at the border. The company must then show that each of its subsequent shipments complies with FDA requirements. To remove itself from an Import Alert, a company generally must show that five consecutive shipments are in compliance. In some instances, a company might also have to submit extensive documentation to satisfy FDA that the company's facilities and operations are in compliance. Import Alerts are posted on FDA's web site, and are therefore readily visible to the general public.

Because FDA can issue an Import Alert based on an *appearance* of adulteration or misbranding, FDA sometimes issues an Import Alert based on evidence that might be vulnerable to challenge. For example, FDA recently attributed a salmonella outbreak to a cantaloupe supplier, and placed the supplier on Import Alert. The supplier sued FDA, alleging that FDA's conclusion regarding the likely source of the outbreak was not rationally supported by the available evidence.[236] If the case had proceeded, the outcome likely would have turned on the sufficiency of the epidemiological evidence relied on by FDA. However, shortly after the lawsuit was filed, FDA rescinded the Import Alert.

In summary, any company that could be subject to adverse publicity in the aftermath of an RFR report, an Import Alert, or other similar situation would be well advised to have in place a plan to manage both the regulatory and the public relations aspects of such an event. Proper management can help minimize the inevitable financial and potential reputational costs, particularly in cases where there might be erroneous attribution.

[235] FDC Act § 801(a).

[236] Del Monte Fresh Produce N.A. v. U.S.A., No. 8:11-cv-02338-JFM (D. MD filed August 22, 2011).

10. Food alert impact and reputation

Jorge Cachinero

10.1 Introduction – the economic importance of reputation

In these times of hardship in the Spanish economy, many of us probably think that it is no time to be thinking about corporate reputation. And considering the basis and hypothesis of the (supposed) acquired value of intangibles in companies' financial and economic capacity, this argument is understandable. Yet in the current times, marked by a declining confidence and credibility of companies and their leaders, the idea has formed and is gaining strength that reputation is not an accessory issue but, on the contrary, something intrinsic in the company's core business and vital in times of economic difficulties.

We are faced with a new macroeconomic and social scenario, which is causing significant, far-reaching changes in companies' management models and their relations with stakeholders. Each according to its size and position, companies face the challenge of dealing not only with a severe financial crisis, but also a crisis of confidence and values.

The adverse economic effect, the lack of leadership, the need to increase levels of transparency and, above all, the lack of confidence has converted analysis, identification and connection with stakeholders into the cornerstone of the reputation management model.

There are very few today who uphold the theory that a company is only responsible for making an economic profit. On the contrary, investigations have produced convincing evidence that companies should seek to achieve gains in five different Profit & Loss (P&L) accounts: economic, governance, environment, people and social.

If we examine each of these, we find that each one is backed by a group of stakeholders: from shareholders to customers, including persons, suppliers and citizens; and each of them requires the development of different conducts, attitudes, commitments and value proposals.

In short, a company with a good reputation has a much better chance in an adverse economic cycle of creating value, establishing better conditions with its suppliers, attracting capital and talent and generating greater customer loyalty. In the long term it will be a more profitable company with a growing number of fans and very few critics.

The Spanish food sector is not, and should not be, oblivious to these ideas even though it is currently in good health and spearheads the Spanish economy. According to the Spanish Federation of Food and Beverage Industries (FIAB), in 2010 the sector accounted for 7.6% of the Spanish GDP with a turnover of €81.369 billion, the third sector with a positive balance of exports and imports only after tourism and the motor vehicle sector.

This makes the food sector, traditionally one of the strongest sectors in Spain, one of the basic pillars for economic recovery. It has an outstanding, widely recognised, exterior projection, which largely favours, or could favour, moreover, an upswing of the reputation of Spain.

Both individually and as a whole, the companies in this sector are well considered in Spain and overseas. A good reputation not only enhances the economic and business yield in normal circumstances, but is also a valuable shield in adverse situations. Its importance within the sector, for example, was demonstrated in the case of the Spanish cucumbers and *Escherichia coli* bacteria crisis in 2011.

The unfounded accusations originating in Germany would probably not have had the same impact, had they been hurled against French products; they may not even have been made. But at that time (and at present), Spain's reputation was not exactly the most adequate or most desirable and, therefore, it did not shield the sector when the crisis broke out.

However, the good reputation of the companies in the food sector has kept them going, maintaining a good pace and health, thanks to their good performance in each of the five P&L accounts mentioned above. Even so, they still have a long way to go.

10.2 The virtuous model of reputation management or how to sail the sea of expressions

The economic importance of one's reputation is clear, therefore, but what is the current general state of the food sector reputation and how can it be improved? Numerous companies in the food sector have experienced situations in the past that have jeopardised their business. However, those with a more sound reputation, with greater support among customers and regulatory authorities and portraying a stronger image as responsible companies have stood their ground much better.

This is the case of Nestlé. In 2010, Greenpeace launched a viral campaign in which it related Nestlé to the destruction of the Indonesian rainforests to enlarge plantations, driving out local communities and destroying the orang-utan's natural habitat. Although admitting the facts, Nestlé's reaction from the first day was to request YouTube to withdraw the English version of the video. But Nestlé's attempt to silence Greenpeace by forcing YouTube to remove the video had a boomerang effect, since hundreds of thousands of internet users watched the video to find out more about the Greenpeace accusation. Nestlé's reputation was exposed to the different communities.

As if this were not enough, Greenpeace published a report, 'The footprint of crime', in which it revealed the process of destroying rainforests in Indonesia to obtain more palm oil. The purpose of the report was to show that the company obtains oil from suppliers such as Sinar Mas, the largest palm oil producer in Indonesia. Greenpeace argued that Sinar Mas was expanding its plantations after cutting down the rainforests and burning and draining the peat-lands, which caused serious social problems, accelerated climate change and destroyed the habitat of endangered species such as the orang-utan.

Greenpeace also stormed the Nestlé Annual General Meeting and supported each of the actions of its activist campaign with slogans such as: 'every time we have a Kit Kat, we are taking a bite out of the Indonesian rainforests' or 'Nestlé should give orang-utans a break and stop using palm oil from suppliers that destroy their habitat, so we also ask consumers to encourage the company to change its procurements policy'.

Hundreds of bloggers from different parts of the world focused their attention on a single supplier. Ikea was also reported by Greenpeace for the same reasons, for the oil it uses in its candles, and in 2008 Greenpeace launched a campaign against Dove cosmetics.

Another case that comes to mind is that of Coca-Cola, but in this case it was not Greenpeace that was behind the campaign but the Belgian minister of health who, without consulting Coca-Cola, announced that the products of the US Company were being taken off the market after finding dozens of cases of poisoning in children. Since the minister gave no details and made no qualifications in his statements, the announcement affected all the varieties of the brand (Coca-Cola, Cherry Coke, Coca-Cola Light, Fanta, Sprite, Sprite Light, Nestea Splash, Nestea, Aquarius, Bon Aqua), causing panic and uncertainty in several countries. Coca-Cola was forced to admit that the problem had been caused by excessive CO_2 in the bottles from the Antwerp factory (northern Belgium) and the presence of fungicides in the pallets used to transport the cans of Coca-Cola manufactured at the Dunkerque plant (France).

Apart from having to convince the authorities and consumers that its products were safe, indicating the source of the raw materials used and the location of its production plants, Coca-Cola had to withdraw all its products from the affected area (Benelux: Belgium, Netherlands and Luxembourg) after finding that the authorities in those countries were recommending people not to consume them.

After seeing its call centres collapse and a loss of several hundred million euros, Coca-Cola saw its reputation take on a far from happy shade.

These cases show us that the reputation of the sector and its companies stems from the state of relations with stakeholders and their direct experience with the company and its employees, and their subsequent opinion of its business activities and value proposal in different dimensions. Deep down, this opinion influences and generates indirect experience in other stakeholders. And nowadays, with the catalytic effect of modern technologies and social networks, that opinion knows no bounds.

In this context, the importance and 'moral' authority of the person expressing his/her opinion are critical elements. And a good understanding, management and efficient use of new technologies are essential for spreading favourable opinions. In short, reputation is the outcome of the perception and opinion held by stakeholders of the commitments that a company has defined in different dimensions, publicly or otherwise, directly or indirectly.

In turn, this is the result of how a company interacts with its critical interlocutors, i.e. how a company has chosen to put across and position the qualities with which it has decided to compete.

This has even greater weight in the food sector, in view of its strong connection with public health, which generates a special sensitivity in all players (regulators, consumers, producers, etc.) participating in it.

But who are the stakeholders and why must the company nurture its relations with them? Stakeholders are any community or interlocutor with moral authority and importance, in the dimensions in which a company (or person, institution, etc.) competes or wishes to compete, that is willing to invest resources (financial or otherwise) in that company's business plan. In other words, they are individuals or organisations that are sufficiently important to be 'listened to' and 'valued' by the environment, who are willing to act or speak positively or negatively about a given company.

Making a brief analysis, companies in the food sector have numerous stakeholders, which we can classify into two groups. On the one hand, there are the traditional stakeholders: customers, individuals, suppliers, partners, society and shareholders or investors (the capital); and on the other, the so-called critical stakeholders, who have acquired moral 'authority' over recent years, personified as pure activists, whose influence and decisions can produce a considerable effect on the smooth running of companies. This group includes: government, third sector, media, bloggers, scientific and academic community and the alumni community of companies and universities.

The major challenge facing organisations in this regard is, on the one hand, to accept that they have lost 'moral' authority and credibility in their environment while other players, in both the digital and analogue worlds, have increased their shares; and on the other hand, to understand that they must win the confidence of their stakeholders and that the latter do not include only investors and shareholders.

They must, therefore, define the necessary tools and processes to build mechanisms of commitment to stakeholders, which are the basis of sustainable development and the levers used by companies to achieve their financial, social, labour, environmental and ethical goals, among others.

In the present context, there are five reasons for believing that an intelligent handling of relations with stakeholders can give organisations a competitive edge: enable a better management of risk and reputation; permit companies and organisations to improve their processes through the transfer of knowledge and the learning of new practices (not everything we do is best); create the channels to meet goals that companies and organisations could not accomplish on their own; substantially increase the ratios of credibility and confidence and, therefore, the respect and admiration for the brands or business of the company; and, in general, lead to fairer, more sustainable social development (Figure 1).

There are two sides to the first step in defining the reputation management model: one theoretical-strategic and the other more ideological-cultural. The first is linked to the analysis of the environment and expressions, opinions and perceptions, which enables companies and/or organisations to find out how confident stakeholders are about a given company and bring the communication strategy in line with the business strategy. The ultimate aim of this stage is to make decisions on the dimensions of reputation in which the organization wants or has to compete and the attributes (differential, naturally) with which it wants to do so. This is very important since it enables us to answer to the following questions: who must I conduct relations with, why and what for? That is where the organisation's strategy lies.

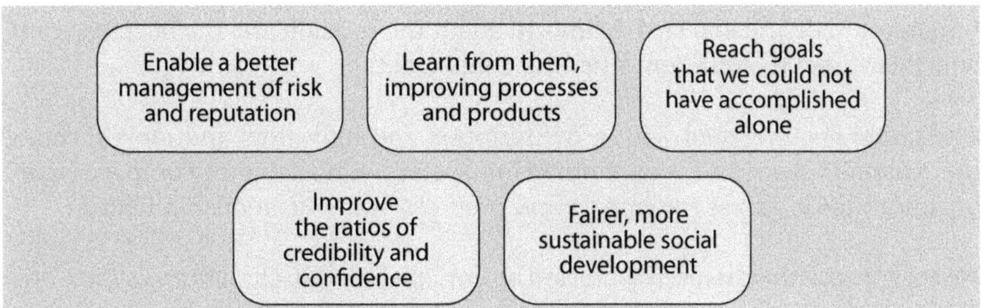

Figure 1. The management of reputation: stakeholder relations.

The second aspect, the ideological-cultural, is concerned with the definition and clear understanding of the company's mission.

There are three ways to address this matter, but the one that best suits the present times and current demands of the sector is the one that indicates, as we mentioned earlier, that 'profit' is but one of the company's goals, along with others: providing a service for people, fostering professional furtherance, protecting the environment, etc. Based on this concept, the mission (or purpose) of the company is explained by its contribution and service to the needs of its different stakeholders. It is, therefore, the basis of the new relationship model for companies. However, in order to understand the company's mission as above, we must overcome the natural tendency to consider profit as something clearly and exclusively of an economic nature.

The definition is also used when telling the company's story. Storytelling builds up the history of the company, highlights its qualities and the differences from its rivals.

After defining the strategy, the next step is to identify the stakeholders (the present ones, which are kept, and the new ones, depending on the decisions made) and classify them according to three criteria of power: relevance, attitude and authority. This subsequently enables us to qualify the 'state' of the relationship and thus prioritise and identify the organisational systems required for effective relationships with each of them.

Behind this, the process of relations with stakeholders, or the Relations and Connection Plan, must be designed. This refers to the programmes, tactics and assets that respond individually to the needs, attributes and strategic issues identified in the first two phases, the ultimate aim of which is to gain credibility/respect/admiration in the surrounding environment and increase the levels of confidence among stakeholders by implementing shared value activities.

Stakeholder management is extremely complex since, as a rule, different groups of stakeholders will have opposing interests and expectations. This is why it is important to define and clearly inform about the mission and the business gain, which is where the different interests come together.

The circle is completed with a definition of the indicators and tools forming the balanced scorecard – or monitoring system – available to companies and organisations to assess the impact and progress of the Relationship Plan.

We suggest that the balanced scorecard be enriched with diachronic and synchronic elements so that the everyday situation can be assessed, taking this to mean the progress and impact of our actions and how they affect the attitude, confidence, commitment and satisfaction of our principal and critical stakeholders.

It will also be necessary – in the medium term – to analyse and assess the impact of the decisions adopted at the initial phase and whether the strategy has enabled the company or organisation to make positive progress in respect of opinions and perceptions.

The foregoing, overall, gives companies and organisations an opportunity to play a more important role, but above all to listen and participate in order to generate a greater economic, environmental and social impact.

As a result of the foregoing, we can see that the value of a company is not always determined by its financial performance, as expressed by certain rankings used in the business world, where, for example, the companies with the highest values on the stock exchange are often not the ones with the largest turnover.

They are, nevertheless, among those most admired by users or customers or, for example, those most sought after as a place to work. This observation corresponds to the studies that show that the 'soft attributes' (or dimensions), those not related with the purely business or economic activity, have a greater overall weight in the valuation of a company than the 'hard attributes' or direct links to economic results (Figure 2).

+ Revenue	+ Value	+ Desired	+ Admired
Exxon Mobile	Apple	Google	Apple
Wal-Mart	Exxon Mobile	BCG	Google
Chevron	Petrochina	SAS Institute	Amazon.com
Conoco Philips	ICBC	Wegmans Foods	Coca-Cola
GM	Wal-Mart	Edward Jones	IBM
General Electric	Microsoft	NetApp	FedEx
Berkshire Hathaway	HSBC	Candem Property	Berkshire Hathaway
Fannie Mae	China Bank	Recreational Equip.	Starbucks
Ford	Petrobas	CHG Healthcare	P&G
HP	Google	Quicken Loans	Southwest
Source: Fortune 500	Source: Media	Source: GPW	Source: Fortune

Figure 2. The economics of reputation: Trust is at the Heart.

10.3 The management of influence on a political level

One of the objectives of public authorities is to protect consumers and users. The Constitutions of different countries contemplate this duty of active observance in one way or another. The food sector is one of the most sensitive sectors for the authorities. Its products are consumed by the entire population and whenever a problem arises, the consequences can be more serious than those encountered in other sectors.

For these reasons, the government will be forced to take sides in any food crisis, make regular statements and adopt decisions. In many cases, political activity will be disrupted by the reality of the crisis.

In 2011, during the outbreak of *E. coli* in Germany, the health authorities of the European Union and its member states often had to make rapid decisions based on false premises that had very serious consequences for certain agricultural products from Spain. In fact, this crisis is still remembered as 'the cucumber crisis', even though in the end Spanish cucumbers were shown not to be responsible for the outbreak.

The pressing need to calm down the population, uncertain as to what food product could be responsible, led to precipitate replies, decisions and political statements. As new cases of *E. coli* patients appeared, measures were taken against Spanish vegetables (not just cucumbers) in Germany, Sweden, Denmark, Finland, the Czech Republic, Belgium, Austria, Russia and the United States.

On 31 May, four days after her initial declarations, Hamburg senator Cornelia Prüfer-Stocks declared that the Spanish cucumbers were not the source of the outbreak, although they did contain *E. coli*. This did nothing to dispel doubts. The next day, the European Commission lifted the health alert.

10.3.1 From the open communication channel...

Since food is a strictly regulated area, politicians' reputations are also at risk in a crisis. This should be borne in mind in all communications with the government.

As mentioned earlier, the authorities always play an extremely important role in any food crisis. Decisions will be made based on the current knowledge of the problem and a calculation of the risks entailed. One must be aware of whether a company is responsible for the situation or just a mere player in the affected sectors.

What information do the authorities receive on what is happening? The sources of information vary widely, but what is essential is that those affected have a significant bearing on the opinion formed by the government.

For this purpose, the communication channel must be permanently open, allowing information to flow freely and transparently. Experience shows that it is not usually a good idea to conceal information. That information will eventually come out and concealing it could break the trust created with the government. Therefore, the greater its responsibility in the cause of the problem, the more proactive and transparent a company should be.

Nevertheless, companies not only have to give information, but should also know to whom they are giving it. It is possible, and can be recommended as we shall see below, that a company may have prior contacts with members of the government. Those contacts are usually on a technical level, but when there is a crisis with public repercussion, the profile of those contacts must be raised. This does not mean that the usual contact with technical advisers must be ignored, but that the company must extend its contacts to encompass those who are going to act directly in respect of the crisis: ministers, secretaries of state, secretaries general, presidents of regional authorities, etc.

Therefore, before any crisis a company should have mapped all the ministries involved and other politicians with interests in the matter.

In Spain or Germany, for example, the powers in relation to matters regarding food and consumption are assigned to the regional authorities (*Länder* or *Comunidades Autónomas*) so the map of strategic players would be extensive, on a horizontal level (MPs, staff of the ministries, regional government authorities) and a vertical level (European, national and regional institutions).

Finally, special care should be taken with the storytelling that is going to be undertaken during the crisis. The message put across must be coherent but the language ought to be adapted to the different interlocutors.

Laboratory researchers, for example, will accept and understand a technical message. On the contrary, depending on the responsibilities of a ministry, it is understandable if a minister or secretary of state does not have any specific technical capacity and needs a different explanation. Moreover, in a crisis a politician will most likely not even have enough time to understand the problem in all its dimensions.

Therefore, when directly addressing a minister or regional councillor, some simple rules should be followed, whether during the crisis, on one or several occasions, or at a personal meeting or simply sending a document by e-mail.

- *Political messages*: politicians will always look for political arguments. Their decisions will always be biased by their respective ideologies so they need to understand why a company's position supports their own storytelling. The resources used must be geared towards explaining the political advantages of making or modifying a decision.

- *The message, at the beginning*. The time a politician can give any interlocutor will be very limited. A one-hour meeting can easily turn into a fifteen-minute encounter. It is perfectly possible that he/she may not spend more than five minutes reading a document sent to him/her. For this reason, the key message must be stated clearly and precisely in the first few minutes, or on the first page.
- *From the problem to the solution*. A politician's desk is usually full of problems to be solved some time before the hypothetical food crisis breaks out. Your proposal might be taken as just another problem and go unnoticed. So proper attention might not be paid to it. The danger of not listening is that politicians may end up making decisions that go against business interests. So companies should always offer a solution to a problem; a simple solution that the politician can use to strengthen his own discourse. The politician is more than likely already aware of the food alert so the source should be explained and, above all, what is being done to solve the problem and what measures will be taken to ensure that it does not happen again.
- *Advanced, but simple, arguments*. When a company goes to these meetings, it must use advanced arguments, although avoiding excessively technical language; otherwise it may feel underrated. The politician himself/herself will often ask to go into more detail about an explanation provided initially.
- *Avoid disputes*. Getting into a dispute does not help to clarify ideas, but rather to increase the confrontation. If there is an opposing stand, one must be attentive to find out why. One may even detect it in the conversation. In other cases political intelligence tools will have to be used.
- *Modesty*. When dealing with institutional officers, one must always be modest. This is something to be borne in mind, especially when the speaker is a company executive not accustomed to this kind of contact. A bad attitude can close all the doors for a company in the future.

10.3.2 The management of influence in political spheres

The crisis provides an opportunity to enter into talks with the government, albeit somewhat hastily. If done properly, it will have discernable advantages for reputation and fluent relations in political spheres. But these relations will be more fluent the greater the benefits to be reaped by both sides. In general, the contacts that might be established with the regulator might be differentiated into three levels: relational, reactive and follow-up.

Within the 'relational' level, several 'win-win' collaboration channels are opened, with the company putting the political interest on a private interest level so that tools will be implemented that are useful for the politician with whom the relationship exists.

Within the 'reactive' level, contact is made with the politician for the first time in the middle of an open issue. Therefore, he/she may well have already received information from other sources, on the basis of which he/she may already have formed a more or less solid opinion. Normally it is unknown what his/her attitude will be on receiving the company's information.

At a 'follow-up' level, a company must participate actively in the political conversation, trying to sway it. A company should go to the corresponding Ministry or the affected members of the Parliament to meet those directly in charge of the issue at stake, or alternatively it should send them the appropriate documents. This constant effort of meeting the decision-makers is one that would provide ongoing benefit in terms of respect of a given issue and the credibility of a well-tended reputation.

At the 'reactive' level, communication channels need to have been created well in advance. At the 'follow-up' one, this channel has to be kept active in order to be able to use it in case of need. At the third level, the 'relational' one, the aim is to establish a two-way relationship. It is necessary to study what tools are best adequate to any given company and what the politicians are expecting from us. The aim is for a company not to be seen merely as requesting information or looking for a solution to its problems, but as providing information that is valid, of high-quality and, above all, useful for performing the political duties of the Regulators involved.

An optimum relationship level is established by anticipating political decisions. Once a crisis is solved, contacts with Regulators should be maintained, since public opinion criticisms of insufficient regulation may later push the Regulators to put forward new and stricter controls in any of the business value-chain. This kind of initiative (either within the Government or within Parliament) tends to begin in an informal political setting, behind the scenes, which a company has to be able to detect and influence so that it is not left out due to lack of access or anticipated information.

This is why it is important to change the pattern in which companies and industries operate within the political realm. In Llorente & Cuenca we have observed a growing need for companies and organisations to adapt their public affairs units to a new political and regulatory environment. In this new scenario, informal conversations are the element that must be grasped to 'move ahead and efficiently sway opinion'.

Tools can be established for this purpose to work on the corporate culture of reporting, relations and anticipation. The decision on what kind of tools are best suited to each organisation is made on the basis of a detailed study focusing on four parameters:

- influence potential;
- internal management of influence;
- resistance to change;
- the public.

Based on those four parameters, any given company will obtain an analysis of its own organisation and the possibilities of bearing an influence upon the political world.

It is necessary to stress some ideas regarding how to exert influence upon political circles:

- It must be a *strategic asset* of the organisation, not just based on personal acquaintances and relationships.
- It must be submitted to *planning* (anticipation).
- It must be possible to *account for it* (reporting, annual report).
- It is a corporate tool to build long-term, win-win *sustainable relations.*
- *It is supported by, and fosters reputation* to create value in companies and organisations.

10.4 Crisis management and online activism

The food industry constantly has to deal with major conflicts, from natural disasters that devastated thousands of harvests in the past to notorious contemporary events, such as the cucumber crisis or BSE (mad cow disease). If we associate the concepts 'crisis' and 'food', we will surely be able to recall countless situations of conflict experienced by companies in the sector. Traditionally these were crises in which society has been witness and participant, the causes of which varied considerably.

In recent years, the food industry has had to cope with problems generated outside the company and others caused by its employees; other situations may have arisen with a specific product or affecting a whole sector; being national or international; being real crises or caused by false rumours; associated with the composition of the product or relating to an indirect problem generated by it; and numerous other possibilities.

This has shaped a complex framework through which it can be sensed that the management of a crisis must contemplate many factors in order to adapt to each case.

10.4.1 The changing pattern of food communication

If we analyse the essence of the industry, consumers have changed in recent decades from 'eating', as the action of physiological satisfaction, to 'feeding', an event involving new factors such as health, nutrition, research or physiological balance. Logically, the pattern of communication has changed at the same pace and progressively over the years. Companies now face more consumers who are more aware, active in communication/opinion and highly exposed to information.

But not only consumers have changed. The environment is also becoming more and more confusing. New players and stakeholders appear every day, interfering with the play of communication. They unquestionably make the scenario increasingly more complex.

As a result, and directly related with the appearance of 2.0 environments, public opinion is now also published opinion. Although this opens up countless new opportunities for companies in terms of marketing and advertising, it also generates a large number of risks, which may develop into a crisis at any time, hampering their management.

Although many companies have documents and manuals preparing them to react in a conventional crisis, nowadays the speed and force of response required makes them insufficient. All this, together with human factors such as subjective decision-making and the tendency to improvise when factors appear related to the online world, can determine the success or failure of the management of these critical situations.

10.4.2 Risk management: secret ingredient of crisis management

Real problems arise unexpectedly. A minor incident can rapidly turn into a crisis with major repercussions on the business and reputation of a company. To avoid this, there are two concepts that must not be overlooked, namely *Preparation* and *Management Tools*.

Both online and offline, being prepared for a possible crisis is essential to stop it from getting any worse. This is why it is necessary to minimise any risks that have been identified and that could set off a serious problem.

The first stage of prevention is preparation. There are some risks one ought to anticipate, by educating consumers, boosting their confidence, informing on progress as it is made and enhancing product information.

Secondly, a company must have tools that enable it to detect risks rapidly and handle any incident immediately and objectively. There are both online and offline tools to make this work easier.

In the offline area, crisis manuals are an example of tools for prevention, since they include messages, procedures, scenarios and recommendations. The latter may be made automatically or adapted to the online environment with handy tools that make them considerably easier to use.

With adequate planning one can prevent a crisis or at least mitigate its most harmful effects.

10.4.3 Managing a crisis in food companies

The line between worlds 1.0 and 2.0 is growing fainter. At present there are very few everyday situations in which the online environment is kept separate from the offline activity and vice versa. For this reason, it is vitally important to have suitable tools to make a combined management of the online and offline environments. And above all, enabling this to be done in the shortest possible time.

There are certain elements in crisis communication that facilitate management of the problem. The digitalization of crisis management contents and material through a darksite or online platform will undoubtedly speed up our reaction. Similarly, having an early alert detection system, such as GUIA designed by Llorente & Cuenca, will enable a company to act swiftly right from the start of the incident.

In the same way, it is important to have a crisis mentality in the organisation, created in advance and enabling management to act in a coordinated and secure way.

The protocols for managing a crisis are similar both online and offline. We describe below the principal phases (Figure 3):

1. *Assessment of the seriousness of the incident* and, more importantly, to do so objectively. To determine how serious an incident is, consider both its source and its effect, which requires appropriate alert and action procedures to handle the incident.
   Something which is not at all ridiculous in an environment in which reactions are not always balanced: people tend to over-react or under-react. Therefore, to cross the barrier from the signs to the crisis requires an analysis enabling a correct diagnosis of the crisis.

Figure 3. The management of reputation: be prepared; worst happens.

2. *Activation of alert protocols and reporting lines*, in other words, coordinating team organisation: establish the profiles, tasks and responsibilities of those involved in crisis management. It is also crucially important to have an early alert system to identify the risks before they actually occur.
3. *Management procedure*: Organising the corporate response process to a reputational risk or crisis involves taking measures online and offline; these may be different. The main keys for this are:
   - Give the right message, in the right measure and to the right target audience.
   - If online communication is necessary, it is best to do so using the company's own tools (corporate networks, website, etc.). What measures are best suited to the situation should be assessed.
   - Humble attitude. Admit proven errors and apologise.
   - It is important to be proactive online. Offline, it depends on the situation.
   - Always explain what is going to be done to solve the problem.
   - In the same way, always explain what the organisation plans to do to prevent similar incidents in the future.
4. *reputation recovery*. When the crisis ends, or starts to wane, it is very important to remain alert and continue taking several actions. We must:
   - Increase the intensity of monitoring and add new search elements, especially online.
   - Implement new scenarios adapted to the new corporate reality after the crisis.
   - Include more 'positive' content in our communications, both through our websites and blogs and offline.
   - Promote the positioning of favourable contents in search engines online (SEO).
   - Assess the development of the crisis, reactions and possible faults that could be avoided in similar events.

10.5 Reputational risks and legal risks

The commencement of any legal action against a company is the beginning of a period in which its reputation will be put to the test in the eyes of all its stakeholders. Almost inevitably, the 'jury of public opinion' will end up, when the legal proceedings have concluded, deciding whether or not to continue trusting the company, its products and its executives. In our experience, that decision is not directly driven by the outcome of the legal proceedings, which may be positive or negative for the company's interests, but depends more on how the company (and other players outside the company) handle communications with each stakeholder throughout the duration of the case.

The food sector has accumulated extensive experience over many years in the management of risks associated with food alerts, product recalls and similar actions. Companies are increasingly more aware that the reputation of their products is put to the test every day on the market (in consumers' purchasing decisions) and have learnt to handle these situations swiftly and efficiently.

However, from the point of view of managing their reputation, the present social and economic scenario is causing additional difficulties for companies in the food sector involved in legal proceedings. The sector faces new challenges, which they will overcome or not depending largely on their capacity to connect with their stakeholders' communication needs and to put across their version credibly and soundly for those who hold the key to their future, from regulators to consumers, including employees, suppliers and distribution chains.

10.5.1 Litigation in the food sector: new stimulants for old risks

It is relatively common for the companies in this sector to face legal actions from consumer associations, which file claims through different channels for a broad array of problems involving quality, labelling, food safety, etc. The sector has also had to fight intense battles over disputes related to anti-trust or fair trading laws, either between producers and manufacturers, or as a sector against national or European anti-trust authorities.

In terms of communication, both these areas (consumer actions and anti-trust disputes) have created major challenges for companies and their brands in the past, since the effect on the purchasing of a product by consumers could be immediate, depending on the sense and significance of the accusations, judgments or resulting penalties imposed on the defendants.

However, two elements stimulating those risks have been triggered recently which have an important effect on reputation management by companies and their brands: on the one hand, increased awareness and sensitivity of public opinion to companies' actions, and on the other hand, the Internet, which has boosted the capacity of mobilisation against companies or their brands and the lasting effects of a case after conclusion of the legal proceedings.

1. *Sensitivity of public opinion to companies' actions.* In the present economic situation, opinion leaders and their followers are more sensitive towards companies that 'do not behave correctly' and may be made scapegoats for the crisis or its consequences. It is becoming increasingly common to observe negative reactions in public opinion to companies that are sued by different players, even leading to protests and demonstrations, boycotts or other forms of activism against their brands and products.

   In recent months, for example, this sensitivity has meant that legal disputes which in other times would have been paid much less attention, such as tax or labour disputes, have become rallying cries against certain brands or companies, even before obtaining a decision in the corresponding legal instance.

   Companies are aware of this situation and no longer consider as 'minor issues' certain conflicts which would have gone unnoticed in other times, or whose public impact would have been limited to a particular group. By way of example, disputes associated with lay-offs sway the public opinion much more strongly in favour of the 'weak' party and against the company, even when there are objective grounds and the correct procedures have been followed. Similarly, the prosecution of legal persons by the Inland Revenue for tax disputes produce a negative reaction among the less specialised public against the companies that appear to be unsupportive of the economic situation, by trying to pay less tax. This scenario makes it essential to prepare very carefully for the management of communication during these proceedings which, as mentioned above, would in other times have had a very small effect on company and product reputations.

2. *Capacity to protest against companies or their brands.* It is widely accepted that the Internet has confirmed itself over recent years as a platform for conversation and the creation of opinion, with speed as its main feature. However, for this analysis, it is more important to bear in mind the capacity it has been developing for some time now as a tool for protesting and calling to action.
   In the food industry, this quality of the different networks we now umbrella under the term 'Internet' or refer to generally as 'social networks' has become an especially sensitive risk for the reputation of a company or its brands and the business, owing to its potential immediate effect on sales and consumer decisions.
   The growing capacity of organisation of 'online activism' forces companies to have an appropriate active listening system on the Internet to be able to react appropriately and in good time to this medium, following an adequate online crisis management strategy.
3. *The challenge of the 'footprint' of a case on the Internet.* The so-called 'right to forget' has arisen precisely from the need to alter the fact that a person or company is 'labelled' with certain attributes, based on true or false information, which have an adverse effect on that person's or company's personal or professional life. It is not difficult to find examples on the Internet of companies or products in the food sector that were immersed in legal proceedings and even years later still appear associated with that fact in the first search results produced by search engines.
   Owing to the characteristics of the Internet, it is not easy to shed the 'millstone' that information from the past (true or otherwise) can come to be, although the desire to do so is perfectly legitimate. It is possible to restore someone's reputation on the Internet and in social networks (reputation recovery) to give a more accurate image of the current situation of a person or company without highlighting certain details from the past by using adequate tools and taking the right actions, which, contrary to popular belief, are not so new in terms of communication because the need has always existed for certain companies to make certain chapters in their past forgotten.

10.5.2 Communication during litigation and its impact on the reputation of a company or brand

As a starting point, we must bear in mind that legal proceedings are a situation of risk for any company, even when it is the plaintiff. Apart from the fact that the climate of information generated on the case will affect the outcome, the way in which communication is handled during the proceedings will affect the company's reputation, regardless of whether judgment is passed in favour of or against it.

All legal disputes involving a company have a strictly legal aspect and another perspective related with the reputational risk. To correctly handle the latter, a communication strategy aligned with the legal strategy must be designed and implemented, especially if the aim is to minimise public impact of the case.

Particularly when at least one of the parties is a well-known person, company or product, litigation is no longer considered a technical dispute within the courtroom but as a competition in which each of the parties must make his version heard, understood and remembered by those having an influential voice in the 'trial by media' and by its key stakeholders.

Experience has shown that it is practically impossible to avoid 'trial by media', especially if the case is well-known or important. What is important is to bear in mind that, to a greater or lesser extent, a case will always arouse a certain amount of interest (by the media, net surfers, etc.), and that, depending on the degree of interest and what is considered best for the company's reputation, each move must be planned considering not only the traditional media but also with our sights set on all the groups or individuals that may be involved or interested in the affair, because the Internet offers them a readily accessible space, simple and absolutely public, in which they can express their opinion and hold a parallel 'trial', even spurring on protests and demonstrations.

The Internet has a bearing on all phases of the judicial process, because it picks up all conversations generated on and about a case and greatly facilitates searches in newspapers and periodicals for a given case at any time. This is a fact that affects and is present in the initial stage and in the processing and judgment of the dispute, offering instant means for assessing each landmark and comparing and discussing it with what society thinks of the case. The Internet participates and plays a leading role in each of these stages, acting as a bridge for those people and institutions with the greatest social influence to express their views on the progress made. The cycle is closed at this point and the group of 'influencers' on each stakeholder (for example, suppliers, employees, customers and regulators) contributes towards creating the context for the following decision or progress in the proceedings. According to this vision, the phenomenon will be repeated in each stage of the proceeding, following a cycle.

In short, public opinion on a court case is no longer formed by a linear mechanism produced as a cause-effect relationship in the traditional media (where each development in the proceedings is commented in the press), but is a cyclical phenomenon in which, after each development in the proceedings and its reflection in the media, new information and opinion channels come into play and sway public opinion, namely the Internet and social networking, leading to a logical effect on those people who, in terms of reputation, we could call 'influencers' (Figure 4).

5 Keys to success in the communication of litigation (Figure 5):

- *Leadership of the legal team* – The communication strategy must always be developed in line with the legal strategy, taking account of objectives, general defence strategy, landmarks and developments of the calendar of the major decisions in each judicial instance, and the principal communication times.
- *Correct diagnosis of the reputational risk* – When analysing the case, specialised tools are needed to detect the media potential of the process. By identifying

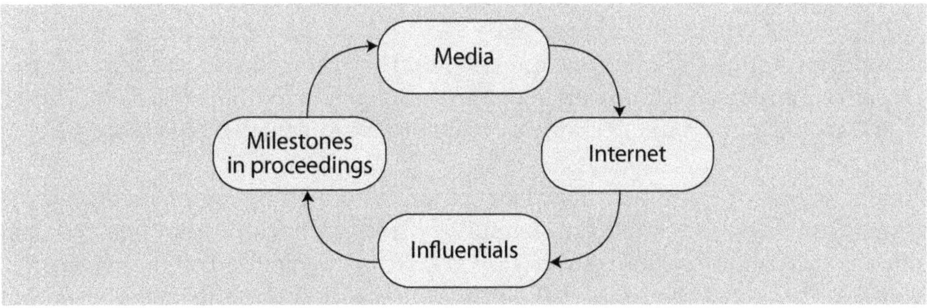

Figure 4. The management of reputation: the ruling of the public opinion.

Social and economical scenario

- Public opinion is more sensitive to companies' actions

The role of internet

- Capacity to raise protests against companies or their brands
- Durability of a case beyond the duration of the legal proceedings

5 keys to success

1. Leadership of the legal team
2. Correct diagnosis of the reputational risk
3. Multi-stakeholder communication strategy
4. Monitoring of the 'trial by media'
5. Training of spokesmen/women

Figure 5. New stimulants for old risks.

the main features of the case and detecting the media case law generated, an initial assessment is made of the case to determine the degree of media interest.
- *Multi-stakeholder communication strategy* – It is vital to have a communication strategy with a multi-stakeholder focus. As mentioned earlier, the reputation of a company, product or person consists of the opinion held of it/him/her by the relevant stakeholders, so if legal proceedings entail a reputational risk, it would be logical to implement a strategy that distinguishes between the different stakeholders and meets the communication needs of each one.
- *Monitoring of the 'trial by media'* – It is important to monitor the information published in the media, both printed and online, and what is posted in the social networks, in order to gauge public opinion, locate the journalists and stakeholders interested in the case, and react and even modify the strategy if necessary.
- *Training of spokesmen/women* – The better the spokesman, i.e. the more convincing and credible he/she is, the greater the possibility that public opinion may take sides with the company's interests in the litigation and, therefore, the stronger its reputation will be, regardless of the final outcome of the court case. Not only must the content of the message be adequate, but also the person communicating it. A good spokesman can play a key role in the outcome of litigation. Although the judges will be guided by their own criteria for interpreting the rules and applying them to the specific case in hand, the public opinion that forms on a legal dispute, partly thanks to declarations by the spokesman, is unquestionably a factor that affects the final outcome.

Finally, it must be stressed that the reputation of a company or product is an aspect that must be treated very carefully when involved in legal proceedings, since it is difficult to avoid a simultaneous 'trial by media'. In the food sector, moreover, there are certain stimulants which require communication during the litigation taking account of all the company's stakeholders with which it wishes to maintain a fruitful long-term relationship of trust.

10.6 Conclusions: be, say and do ... and change your business model

The model of relationships between the company and its stakeholders is changing and the search for, interpretation and care of their interests are of vital importance. The business environment is growing more and more demanding and companies are required to meet unprecedented levels of transparency and visibility. This situation has been heightened by the global nature of companies, mainly due to the astounding development of information technologies, which has increased social scrutiny and their exposure to the entire planet.

Some of the challenges facing companies are, therefore: the generation of empathy to transmit confidence and credibility to the critical stakeholders; astuteness to develop customised connection proposals in order to influence them; and a territorial vision to handle an increasingly more global, integrated and controlled environment, in which the extra ingredient is to prove that they are committed to culture, values and the society to which they belong and in which they perform or wish to perform their business activities. This requires a new focus on reputation and management methods and tools that convert hitherto mechanical processes, which have become obsolete, into a true particle accelerator.

The company must, therefore, identify the business dimensions in which it wants to compete; design the story, formed by attributes and strategic intentions, with which it wants to compete; analyse and establish a hierarchy for the critical issues of stakeholders; design connection proposals to influence critical stakeholders; and set up a measurement system to enhance the value of its intangibles. In other words, it must, through a strategic relationship, connection and permanent dialogue with stakeholders, create a competitive edge per se that enables the company and its management team to be sufficiently outstanding to compete through differentiation.

We are convinced that the management of intangibles through the reputation model and connection with stakeholders creates in itself a shield against crises and emerges as a lever for creating value to enable the company to compete on the markets with the approval of those having power.

11. EU competition law as applied in the agriculture sector

Nicole Coutrelis

During the past recent years, the very structure of the Common Agricultural Policy (CAP) has been progressively modified, coming from the 'old' times when almost all prices and market management was harmonised and decided at the EU level, to the current system embedded in the last version of the 'Single Common Market Organisation' (Single CMO) in Regulation (EU) No 1308/2013 of 17 December 2013, which entered into application on 1 January 2014.

The current system is far less regulated than it used to be, and in many sectors common prices and common management of the market have disappeared, replaced by direct subsidies to farmers which are no longer coupled with their production.

In parallel to this dramatic change, competition rules have also changed, but they still are specific to the agriculture sector. Therefore, as long as agricultural products are concerned, it is always necessary to check how competition rules apply to them in a particular situation.

11.1 The basic principles of EU competition law

- Article 101 of the Treaty on the Functioning of the European Union (TFEU) prohibits agreements that could disrupt free competition in the European Economic Area's internal market.
- Article 101(1) provides that: The following shall be prohibited as incompatible with the internal market all agreements between undertakings, decisions by associations of undertakings and concerted practices which may affect trade between Member States and which have as their object or effect the prevention, restriction or distortion of competition within the internal market, and in particular those which:
  - directly or indirectly fix purchase or selling prices or any other trading conditions;
  - limit or control production, markets, technical development, or investment;
  - share markets or sources of supply;
  - apply dissimilar conditions to equivalent transactions with other trading parties, thereby placing them at a competitive disadvantage;
  - make the conclusion of contracts subject to acceptance by the other parties of supplementary obligations which, by their nature or according to commercial usage, have no connection with the subject of such contracts.

- Article 101(2) provides for the nullity of agreements taken in infringement to article 101(1).
- Article 101(3) provides for possible exemptions from article 101(1), in favour of agreements considered to have positive effects without totally excluding competition, and which provide benefits to consumers.

Since 2004 (entry into application of Council Regulation (EC) No 1/2003 'on the implementation of the rules on competition laid down in Articles 81 and 82 of the Treaty'[237]), Article 101(3) is directly applicable and the Commission does not grant any more individual exemption decisions as it used to do in the past under Regulation 17. It is now the responsibility of the parties to an agreement to evaluate whether this agreement meets the required conditions for an exemption.[238] Arguments based on Article 101(3) are to be raised only in litigation when agreements are challenged by the Commission or by national competition authorities, or before national courts in private litigation.

Article 102 TFEU prohibits the abuse of a dominant position, which may be committed by a single undertaking or by several jointly (collective abuse):

> Any abuse by one or more undertakings of a dominant position within the internal market or in a substantial part of it shall be prohibited as incompatible with the internal market in so far as it may affect trade between Member States.
> Such abuse may, in particular, consist in:
> 1. Directly or indirectly imposing unfair purchase or selling prices or other unfair trading conditions.
> 2. Limiting production, markets or technical development to the prejudice of consumers.
> 3. Applying dissimilar conditions to equivalent transactions with other trading parties, thereby placing them at a competitive disadvantage.
> 4. Making the conclusion of contracts subject to acceptance by the other parties of supplementary obligations which, by their nature or according to commercial usage, have no connection with the subject of such contracts.

No exemption is provided concerning abuse of a dominant position: it is up to the judge, or to competition authorities, to assess whether an undertaking (or a group of undertakings) enjoys a dominant position on a specific market and if yes, whether its conduct is abusive within the meaning of Article 102.

---

[237] Articles 81 and 82 are now replaced by Articles 101 and 102, respectively. For sake of clarity, we will use these current numbers, even for 'old' regulations which referred to old Treaty numbers.

[238] The 'block exemption' regulations, for instance for distribution (vertical agreements), or for some cooperation agreements (horizontal agreements), are not addressed in this chapter, which is focused on the specific regime of agricultural products, although they may apply also when agricultural products are concerned.

As to agricultural products (i.e. products listed in Annex I TFEU), these rules do not apply directly, but through specific provisions of the Treaty.

11.2 The specific application of competition rules in the agriculture sector: a brief historical overview

According to *Article 42* TFEU:

> The provisions of the Chapter relating to rules on competition shall apply to production of and trade in agricultural products only to the extent determined by the European Parliament and the Council within the framework of Article 43(2) and in accordance with the procedure laid down therein, account being taken of the objectives set out in Article 39.[239]

Therefore, in order to analyse competition rules applicable to a particular agricultural sector, reference should be made to the specific Parliament and Council regulation taken to this purpose.[240]

As far as antitrust is concerned, this provision of the Treaty has for many years been enforced by a general Council Regulation taken in 1962, Regulation 26 'applying certain rules of competition to production of and trade in agricultural products', which has been replaced only in 2006 by Council Regulation (EC) No 1184/2006 of 24 July 2006.

Regulation 26 was a horizontal regulation which covered all agricultural products. To sum up very broadly: It stated, as a principle, that competition rules were applicable to agricultural products. However, some derogations to the applicability of Article 101 were laid down.

In law, this means that when the specific derogations are not applicable, Article 101 is applicable in its entirety (including possible exemptions under Article 101(3)). No exception has ever been laid down for Article 102.

As to Article 101(1) – and consequently the nullity provided for in Article 101(2) and the necessity for an exemption under Article 101(3) -, it did not apply to agreements and practices

- which were part of national market organisations, and
- to those which were 'necessary for the attainment of the objectives of Article 39' (objectives of the CAP).

239 This provision covers antitrust (Articles 101 and 102) as well as State aid (Articles 107 – 108). The present chapter does not address State aid, but only antitrust.

240 Before the entry into force of the co-decision procedure, Article 42 of the Rome treaty referred solely to the 'Council', without reference to the Parliament.

- A specific provision was devoted to horizontal agreements between farmers 'belonging to a single Member State which concern the production or sale of agricultural products or the use of joint facilities for the storage, treatment or processing of agricultural products, and under which there is no obligation to charge identical prices, unless the Commission finds that competition is thereby excluded or that the objectives of Article 39 of the Treaty are jeopardized'.

An important procedural provisions also stated that the Commission had sole competence (under the supervision of the ECJ) to decide whether such conditions for non-applicability of Article 101 were met.

A long standing practice of the Commission, backed by the jurisprudence of the ECJ, has from the very start established that the exceptions to the applicability of Article 101 laid down in Regulation 26 had to be construed very narrowly. First, because, like any exception, it is a common principle of law that they should be interpreted strictly. But also, because the very system lead necessarily to a very restrictive application. Indeed, it appeared very soon, that:

- Very few national market organisation remained, since almost all agriculture was progressively covered by Common Market Organisations (CMOs);[241]
- When there was a CMO in one sector, it was supposed to be sufficient to reach the objectives of Article 39. So, it left no competence to national authorities, neither for any 'national market organization', nor *a fortiori* for any measure taken by private operators under this 'umbrella'. The reasoning was in fact very simple:
  - either the CMO covered the same purpose as the agreement at stake (for instance, product standards, or price mechanism), and then it was not possible to take different measures than those provided for in the CMO;
  - or the CMO did not rule this matter, and this meant, in general, that such measure had not been considered to be 'necessary for the attainment of the objectives' of the CAP.
- In addition, and assuming (for sake of reasoning), that a measure could be envisaged without falling into these prohibitions, it was almost impossible to prove that a private agreement could be necessary to attain all the objectives of Article 39 at the same time.

So, in practice, and save for very few exceptions, the only agreements between undertakings in the agricultural sector which were not subject to Article 101 were those which were regulated in the very CMO governing the sector to which the concerned products belonged.

[241] The only significant sector which has remained out of scope of any common market organisation is the potato sector.

For instance, in the '80s, the existence of Producer Organisations, with specific powers to concentrate the placing of products on the market, was introduced in the CMO for fruit and vegetables. Later on, interbranch organisations (i.e. vertical agreements between farmers organisations and representatives of downstream operators such as industry or trade) were also recognised under certain conditions.

However, in no case was it admitted that such agreements and organisations could fix prices - something, which was the sole competence of the EU in CMOs (when such prices were fixed).

This evolution, which has given birth to specific competition rules which clearly derogate to the strict application of Article 101 in specific product sectors, has been legally enacted in Regulation (EC) No1184/2006, which has replaced Regulation 26 in 2006. Indeed, Regulation (EC) No 1184/2006 contains the same general rules as Regulation 26 (commented above), but according to its Article 1, it is not applicable to products covered by the Single CMO.[242]

So now, *de facto* and save for a very limited range of products, competition rules as applied to agricultural products are laid down in Regulation (EU) No 1308/2013. The next part of this chapter focuses on the provisions of this regulation

11.3 Current situation: provisions of the single CMO

Regulation (EU) No1308/2013 contains two relevant ranges of provisions:
- General rules regarding competition law applying to all undertakings in the agriculture sector (here after 3.1.).
- Specific rules governing specific sectors (here after 3.2.).

In both cases, when an agreement or a collective decision is found not to be subject to the application of Article 101 TFEU, this entire article is not applicable. This means that the question of potential exemption under Article 101(3) is not relevant. On the other hand, if an agreement or a decision is not covered by the exceptions specifically laid down for the agriculture sector, an exemption under Article 101(3) may still be envisaged in law (even if, in practice, it is generally very difficult to obtain).

[242] The Single CMO, covering all products instead of all the different single CMOs which were in force previously, was first put in place by Regulation1234/2006. In the beginning, this regulation did not cover all products (for instance, wine was included later). Now, the current Single CMO is in Regulation 1308/2013, which covers all regulated products.

11.3.1 General rules regarding competition law applying to all undertakings in the agriculture sector: Articles 206-210 and 222 of Regulation (EU) No1308/2013

Article 206 of the Single CMO lays down as a principle that Articles 101 to 106 TFEU are applicable to the production and trade of agricultural products, 'save as otherwise provided in this regulation'.

The application of the competition rules of the Treaty being the principle, the exceptions must be construed strictly. It stems from Article 206 that these exceptions can only be found in Regulation (EU) No 1308/2013 itself (or eventually in regulations or decisions based on it).

Basic general exceptions are found in Articles 209 and 210 and articles to which they refer. Other exceptions are in sectorial provisions.

Also, according to Article 222, other measures to the effect that Article 101(1) is not applicable may be adopted for any product by the Commission, if necessary 'during periods of severe imbalance in markets', for limited periods of time and under strict conditions. By definition, such measures will be exceptional and very strictly framed, to be taken only after all the other tools existing in the CMO have been taken to manage the crisis. The potential application of this article is not explored further in the present chapter. Let us just have in mind that, even more than for any other measure derogating to competition rules, they will have to be construed strictly.

11.3.1.1 Article 209: agreements necessary for the attainment of the objectives of the CAP and agreements between farmers and their associations

Article 209 contains similar provisions to those mentioned above which were in Regulation 26 and are still in Regulation (EC) No 1184/2006, except those referring to national market organisations which, by definition, cannot concern products covered by the Single CMO. Therefore, two basic derogations are still in force: the agreements which are necessary for the attainment of the objectives of the CAP, and some agreements between farmers, provided that they do not exclude competition and that they do not entail an obligation to charge identical prices.

However, two main differences with the 'old' regulations are to be underlined:
- A very important procedural difference: Article 209(2) specifies that 'Agreements, decisions and concerted practices which fulfil the conditions referred to in Paragraph 1 of this Article shall not be prohibited, no prior decision to that effect being required.(...)'. So, it is now up to the operators to make their own evaluation of the compatibility of their agreement with Article 209(1), a situation which does not provide the same legal certainty as previously.

- Much more detailed provisions than previously regarding horizontal agreements between farmers, by reference to producer organisations. Producer organisations are not the only form of horizontal agreements between farmers which benefit from the non applicability of article 101. All forms of agreements found in Regulation 26 are still mentioned, and consequently the 'old' jurisprudence related to these provisions is still valid. However, in addition, producer organisations, which are a specific and quite structured form of horizontal cooperation between farmers, are now part of the 'general rules' regarding competition law in agriculture, whereas they were previously part of specific product CMOs (originally in the fruit and vegetable sector).[243]

As to Producer organisations (POs) one has to refer to Article 152 of Regulation 1308/2013. These POs can be formally recognised by Member States, and can be set up in any product sector. Their objectives can be very diverse, and can go as far as 'the concentration of supply and placing on the market the products of their members' (Article 152(1)(c)(ii)). However, except when they are explicitly authorised to negotiate prices on behalf of their members (see below in 3.2.), they cannot take any measure as to prices.

Also and in any case, Article 209(1) does not allow producer organisations to make agreements between themselves or with independent growers to fix a common sale price.

11.3.1.2 Article 210: agreements and concerted practices of recognised interbranch organisations

Article 210 provides that, under certain conditions, Article 101 TFEU is not applicable to certain agreements, decisions and concerted practices of interbranch organisations which have been recognised by Member States. For the application of this article, reference should be made to Articles 157 and 158 of Regulation 1308/2013.

Interbranch organisations are defined in Article 157 as vertical organisations composed of farmers and other downstream operators (mainly, industry and/or trade). These provisions, which come from recent amendments in some sectors, are new as a general derogation to competition rules which may now apply in all sectors.

As opposed to the exceptions provided in Article 209, the one laid down in Article 210 is subject to the notification of the agreements to the Commission, which may oppose to them if it considers that the conditions of the exception are not met

[243] We do not address here producer organisations which were also introduced in the fishery sector in the '80s.

and that, consequently, Article 101 TFEU does apply to that agreement. Such a procedure grants a much greater legal certainty than article 209.

Member States have the obligation to notify to the Commission all interbranch organisations they have recognised (Article 158(5)(e)). They must also carry out regular controls on the activities of these organisations, with a view to withdrawing recognition in case of non-compliance or irregularities (Article 158(5)(b), (c) and (d).

Among the possible aims of an interbranch organisation, Article 157(1)(c)(i) mentions 'improving knowledge and the transparency of production and the market, including by publication of aggregated statistical data on production costs, prices, including where appropriate, prices indices, volumes (...)'. It is therefore possible, within an interbranch organisation, to closely monitor actual costs and prices, information which gives the operators important indication as to how to negotiate and fix their prices. But this provision does not go as far as allowing the very fixation of prices.

In this respect, Article 210(4)(d)) states that agreements, decisions and concerted practices concluded within interbranch organisations 'shall in any case be declared incompatible with Union rules if they (...) entail the fixing of price or the fixing of quotas'. In addition to these general rules which are applicable to all product sectors covered by the Single CMO, the regulation contains numerous specific sectorial provisions, which offer generally more possibilities to derogate from competition rules than the general rules summarised above.

11.3.2 Specific sectorial provisions of Regulation (EU) No 308/2013

Inherited from the former different CMOs which regulated separately every single product sector, various provisions which derogate from competition rules are now found in different parts of the Single CMO. We do not attempt here to expose the detailed provisions governing every single product sector, but rather to give a brief overview of the types of measures which can be found. Needless to say that, in each single case, one has to refer very carefully to all the relevant provisions of the Single CMO, i.e. (1) to the general provisions exposed above; and (2) to the sectorial provisions specific to its product sector.

11.3.2.1 Agreements on prices

Some sectorial derogations to competition rules go as far as allowing collective negotiations of prices by producers organisations on behalf of their members: this is the case for milk (Article 149), for olive oil (Article 169), for cattle in the beef and veal sector (Article 170) and for some arable crops (Article 171).

When such negotiations by producer organisations are authorised, they are strictly framed and are, particularly, limited in terms of market share[244]. They are also taken under the scrutiny of the Commission if they can affect trade between Member States, or of the National Competition Authorities (NCAs) when they have only a national scope.

However, even within such strict limits and under such scrutiny, these derogations deserve to be underlined, because they do allow what is normally considered, in competition law, as a 'hardcore' restriction, which generally cannot benefit from any exemption under Article 101(3).

11.3.2.2 Restrictions of production

Restrictions of production are also generally considered as 'hardcore' restrictions in competition law. Under the Single CMO, Member States are allowed, in some sectors, to take measures restricting the production or the placing of products on the market. This is the case in the wine sector (vine planting rights, Articles 63 and 65), for cheese with a geographic name (Article 150), and for ham with a geographic name (Article 172). What is remarkable regarding those measures if that the professional organisations (producer organisations or interbranch organisations) are associated with the State for the decision.

Indeed, for new vine planting rights, a Member State 'may take into account recommendations presented by recognised professional organisations operating in the wine sector (...) provided that those recommendations are preceded by an agreement entered into by the relevant representative parties in the reference geographical area' (Article 65). In other words, operators on one market have their say regarding the entry on the market of new operators or on the increase of production of other operators, and member States authorities are encouraged to take the views (and interests...) of these operators into account.

For cheese and ham protected by a geographic name, the influence of the producers organisations is even stronger for limiting the production: upon request of professional organisations, the Member States may 'lay down, for a limited period of time, binding rules for the regulation of the supply (...)'. For cheese as well as for ham, the professional organisations must represent at least 2/3 of the producers referred to in the geographical area. In general competition law, this means that producers holding a collective dominant position on a market can ask national authorities to limit the supply, something which is clearly unacceptable

[244] For milk, a producer organisation is authorised to negotiate on behalf of its members only if it does not hold more than 33% of the national market. This maximum share is 15% for arable crops.

under Article 102 TFEU.[245] These measures are strictly framed, and must be notified to the Commission which may, eventually, repeal the measures. It is also clearly stated that such measures cannot allow for price fixing.

Nevertheless, even within such limits, this influence of producers' organisations in the decision making process is remarkable and clearly in derogation from competition law in general. In fact, if they were not expressly authorised by the Single CMO, such measures would clearly infringe Article 4 TEU[246] in conjunction with Articles 101 and 102 TFEU.

11.3.2.3 Various other measures

Without pretending to be exhaustive, we can also mention here some sectorial measures which, if they were not provided for in the regulation, would not be authorised under competition law – or at least would need close scrutiny from the competition authority. This is the case for example with:
- 'Agreements within the trade' to frame some aspects of the contracts between beet growers and sugar manufacturer (although such collective agreements can include neither the fixation of prices nor any limitation of production)
- Possible extension of rules and compulsory contributions to professional organisations, which can be imposed in any sector for some measures such as, for example, the protection of the environment or the definition of minimum qualities. Among the measures which can be rendered compulsory (see Article 164), not all of them are restrictive of competition, but some of them clearly are. This is the case, for instance, with the definition of minimum quality and minimum standards for packing and presentation. Here again, we are clearly facing measures which would not have been acceptable in competition law without such explicit derogation.

As to standards, this provision of the Single CMO is worth to be compared with the general system of the 'new approach' in the industry. In this system, conformity to standards can be taken as references to prove the conformity of products to essential requirements (such requirements being compulsory), but these standards are only optional references. An operator can always try to prove the conformity of his products to the essential requirements by other means than following the official standards. Under the Single CMO, quality standards fixed by professional organisations can be rendered compulsory by Member States authorities, without allowing a non-member of said organisation to diverge from the standard.

[245] Although explicit derogations to antitrust rules seem to be limited to Article 101, we have here a derogation which, de facto, is also contrary to Article 102.
[246] Provisions of former Article 5 EEC, then 10 EC, which prohibit Member States from acting against the EU principles, are now in the Treaty on the European Union (TEU).

11.4 To conclude: a few comments

As opposed to the situation which prevailed before the Single CMO, all rules regarding competition for products subject to the CMO are now in the same text which, although quite complex, can be apprehended with some logic. In our view, the logic starts from Article 206, which is the point from which all the rest derives. According to this article, competition rules (and particularly Articles 101 and 102 TFEU) apply to agricultural products 'save as otherwise provided in this Regulation'. It is therefore clear, in law, that all the exceptions are in Regulation (EU) No1308/2013 (or in texts based on it), and that these exceptions (summarised in the present chapter), like all exceptions, should be construed restrictively.

That said, more is to come. Article 206 also says that the Commission and the National Competition Authorities (NCAs) must apply the competition rules in close cooperation, and that the Commission 'shall, where appropriate' publish guidelines to assist the NCAs. Such guidelines have not been published yet and obviously, this has to be closely monitored.

But even more importantly, it is essential to look at what is happening in the Member States, with or without Commission guidelines. Since 2004, the enforcement of competition law if mostly 'decentralised' within the Member States. Many decisions are taken, not by courts, but by NCAs which may impose fines – and indeed they do. So, more than ever, operators in the agriculture sector must be aware of competition law: anything which is not expressly covered by the derogations found in Regulation (EU) No 1308/2013 is not just a question of interpretation of the regulation, but can be subject to procedures before the NCAs.

On a more political ground, the following few comments come also to mind when analysing the derogations found in Regulation 1308/2013.

Actually, many of these derogation go through producer organisations or interbranch organisations, as if, de facto, these organisations had received some sort of 'delegated powers' to organise and manage the markets in place of the public authorities (mainly EU authorities) in the 'old times' of the first CAP.

Many of these organisations have advocated for years to obtain such powers, but they have to realise that the 'price' to pay for that is a much greater exposure to competition law – something to which all of them were not necessarily prepared. This situation is all the more delicate now that, as opposed to the system which prevailed with Regulations 26 and 1184/2006, decisions taken by producer organisations and, more generally, actions which may be covered by Article 209 are not any more subject to prior approval from the Commission.

It should also be borne in mind that, under Article 101(2) TFEU, agreements and decisions taken in infringement of competition law are null and void. This provision is directly applicable before national courts and may therefore apply in private disputes, even in the absence of any action by NCAs.

Obviously, these specific competition rules found in the Single CMO are linked to the decrease of the protection which was granted by the 'old' CAP: a specific regime for agriculture is probably more useful now than it used to be in the past. This may look paradoxical in a time when the rationale for the successive CAP reforms have always been to 'liberalise' the sector, but this is really understandable considering the intrinsic specificity of agriculture (whose parameters are not in the hands of the operators), and also the necessity to gradually step out of the old CAP. Formerly, all that was not in a CMO was deemed unnecessary and therefore out of the possible derogations from competition rules. The legal reasoning has not changed (all derogations to competition rules must be expressly authorised), but now that some protective rules have disappeared from the CMOs, authorised derogations have increased.

As if the decrease in public regulations has increased 'professional' regulations – much more than really open the sector to real liberalisation. However, it is far too early to draw a final conclusion: the Commission and, probably, more importantly, the NCAs, will play now a considerable role in framing the 'new' CAP.

12. Legal and judicial framework in Italy

The new Italian regulation of commercial transactions in the food sector

Alessandro Artom

12.1 Introduction

On the 24th of March 2012, the Law 24.3.2012 n. 27 relating to 'Urgent provisions on competition, infrastructures development and competitiveness', implementing the so-called *Decreto Liberalizzazioni* or *Cresci Italia*, Decree 24.1.2012, n. 1, has been published in the Italian Official Journal (24.3.2012, n. 71). As everybody in Italy knows, the new regulation is part of a bigger operation arranged by the technical government led by Prime Minister Monti for the growth and recovery of Italy from the economic crisis effects and comprehensive of two other Decrees: Decree 'Salva-Italia' (D. L. 6.12.2011, n. 201, converted by L. 22.12.2011, n. 214) and Decree *Semplificazioni* (D.L. 9.2.2012, n. 5, converted by L. 4.4.2012, n. 35).

Article 62 of the Decree[247] introduces a new regulation, proposed by the new Minister of Agriculture Mario Catania, on commercial contracts in the food sector. On the same issue, in 2007 the draft regulation AS 1644, the so-called 'Bersani-ter', that never came into force, proposed Article 3-bis concerning Antitrust Authority's check on the contractual provisions related to the selling of products to the Retail/ Large Distribution, aimed to avoid any direct or indirect burden for the supplier[248]. The issue relating to the commercial relationships between the consumer goods industry and the Large Distribution has been treated also by the Tax Authority with the Resolution n. 36/E of 7.2.2008, in which the commercial practices in the food sector with reference to the deductible costs have been analysed.[249]

In 2008 commercial contracts in the food sector have constituted the main subject of the Second Panel-Discussion of the Italian Food Law Association (AIDA) on *I contratti del mercato-agroalimentare* (Food sector contracts), held during CIBUS[250]

[247] Decreto Legge, testo coordinato 24.01.2012 n° 1, G.U. 24.03.2012.

[248] Article 3-bis Contractual terms on improper burdens for the suppliers. *'L'Autorità Garante della concorrenza e del mercato vigila e verifica, anche su segnalazione delle associazioni degli imprenditori, che la previsione di clausole contrattuali recanti oneri, diretti o indiretti, a carico del contraente per il suo inserimento nella lista dei fornitori o per l'accesso dei propri prodotti all'esposizione negli esercizi commerciali appartenenti a catene distributive, nonché di altre tipologie di vendita promozionale comunque denominate, non configuri abuso di posizione dominante o pratica anticoncorrenziale ai sensi della legge 10 ottobre 1990, n. 287'.*

[249] Risoluzione n. 36/E del 7 febbraio 2008 dell'Agenzia delle Entrate – Direzione Centrale Normativa e Contenziosa, available at: http://tinyurl.com/q84r4pc.

[250] Parma International Food Exhibition.

2008, where the participants discussed the contracts between farm and food industry, food industry and Large Distribution, contracts with the consumer[251].

Finally, these subjects have also involved the Italian Antitrust Authority, which on the 27th of October 2010 started a research on the role of the Retail/Large Distribution within the entire agricultural and food chain, with particular reference to: the effective level of competition existing between food companies, that are part of different food sector groups; the marketing contractual trends and the eventual competitive relevance, also in the industrial production market, of the behaviours adopted by Large Distribution's operators during the negotiation of the purchasing conditions.[252]

12.2 Scope and essential elements of the contract – Article 62(1)

The first paragraph of Article 62 defines the scope of the new provision, which has to be considered applicable to all transactions or contracts of sale of agricultural products and foodstuff as defined by Regulation (EC) No 178/2002:

> any substance or product, whether processed, partially processed or unprocessed, intended to be, or reasonably expected to be ingested by humans. 'Food' includes drink, chewing gum and any substance, including water, intentionally incorporated into the food during its production, preparation or treatment[253].

The reference to the European provision is confirmed by the definition of foodstuff provided by Article 2(b) of the draft inter-ministerial decree of the Minister of Agriculture and the Minister of Economic Development adopted in accordance to Paragraph 11 bis of Article 62, that makes reference to the definition of Article 2 of Regulation (EC) No 178/2002. The contracts covered by the scope of the new provisions are the ones signed by 'practitioners',[254] excluding the contracts/sales to the final consumer.

---

[251] See the article in the on line Legal Journal – Rivista di diritto alimentare, Anno II, numero 2 – Aprile-Giugno 2008 su www.rivistadirittoalimentare.it.

[252] See the article of Alessandro Artom, *'L'indagine conoscitiva dell'AGCM sul ruolo della GDO nella distribuzione agro-alimentare'*, in the on line Legal Journal – Rivista di diritto alimentare (Anno IV, numero 3 – Luglio-Settembre 2010) at www.rivistadirittoalimentare.it.

[253] Article 2 Regulation (EC) No 178/2002 of 28 January 2002.

[254] 'Practitioner' as defined by Article 3(c) of the Italian Consumer Code (D.Lgs. 6 settembre 2005, n. 206 Codice del Consumo): *'professionista: la persona fisica o giuridica che agisce nell'esercizio della propria attività imprenditoriale, commerciale, artigianale o professionale, ovvero un suo intermediario'.*

Concerning the scope of Article 62, the implementing decree mentioned above specifies in Article 1 that Article 62 is applicable to the commercial transactions in the food sector, delivered on the Italian Republic territory.

Paragraph 2 of the same Article 1 does not include into the scope of Article 62 three categories of contracts:
1. orders of foodstuff and agricultural products made by companies to the cooperatives of Article 2(2) of Decree 228/2001, when that companies are associated to that cooperatives;
2. orders of foodstuff and agricultural products made by companies to the producers' organisations of Decree 102/2005, when those companies are associated to those organisations;
3. orders of fish products made by fish companies of Article 4 of Decree 4/2012.

It is also specified that the scope of Article 62, Paragraphs 1 and 3 does not include simultaneous transactions with delivery and payment happening at the same time (i.e. sales to consumers).

In general, the contractual relationships between food industry and Large Distribution concern two cases: (1) foodstuff supply contracts between food industry and Large Distribution (the latter considered individually or as central buying office) with periodical commitment; (2) services provided by Large Distribution/customer to the supplier/foodstuff producer.

We deem that among the practitioners it is possible to include also the Public Administration, through the analogical application of Article 2(1)(a) of Decree 231/2002 on late payments in commercial transactions (this provision is referenced indirectly by Paragraph 3 of Article 62, and directly by Paragraph 11), where commercial relationships are defined: 'any contract, no matter how defined, between companies or between companies and Public Administration that causes the delivery of goods on payment of price'[255].

The new regulation establishes that these contracts have to be made compulsorily in written form and have to contain determined elements, the lack of which leads to the nullity of the contracts: 'duration, quantities, features of the product to sell, price, terms of delivery and payment'[256].

[255] D.Lgs. 9 ottobre 2002, n. 231 Implementation of Directive 2000/35/EC of the European Parliament and of the Council of 29 June 2000 on combating late payment in commercial transactions. Article 2(a) transazioni commerciali (commercial transactions): *'i contratti comunque denominati tra imprese ovvero tra imprese e pubbliche amministrazioni che comportano in via esclusiva o prevalente la consegna di merci contro il pagamento di un prezzo'.*

[256] Article 62(1): '... *la durata, le quantità e le caratteristiche del prodotto venduto, il prezzo, le modalità di consegna e di pagamento'.*

The implementing Decree specifies in Article 3(2) when the required written form is satisfied for the purpose of the Decree. In particular, it suffices to have:

> any form of written communication, even electronic or by fax, even without signature, with the aim to show the parties' willingness to constitute, regulate or extinguish a commercial relation concerning transactions or contracts of sale of the products of Article 2(a) and (b) (agricultural products and foodstuff).

According to Article 3(3) of the implementing Decree, the essential elements of the contracts have to be made in written form as well and can be contained both in the contracts or agreements of Article 2(1)(l) and (m) of the Decree (framework contracts; basic contracts; inter-professional agreements) and in the following documents, provided they contain the essential data and the reference to the corresponding contracts or agreements:
1. contracts of sale of products;
2. transport or delivery documents, or the invoice;
3. purchase order by which the purchaser commits the products' delivery.

The essential elements, in written form, can also be contained in exchange of communications and orders, previous to the products' delivery (Article 3(4)). The transport or delivery documents, as well as the invoices, integrated with all the elements required by Article 62(1) (except for the cases of Article 3(3)) fulfil the obligations of Paragraph 1 and have to report the following wording: 'fulfils the obligations of Article 62, Paragraph 1 of the Decree 24th January 2012, n. 1, converted with modification by Law 24th March 2012, n. 27'.

It is important to underline that the fact that the provision identifies the written form as compulsory includes these contracts in the category of acts 'specifically indicated by law' foreseen by n. 13 of Article 1350 of Italian Civil Code, establishing that these acts has to be made in a particular form, that is to say public record or private deed. As a consequence, the form indicated by Article 62 is a form *ad substantiam* and therefore, it is an essential requirement according to Article 1325 n. 4 of the Civil Code, the lack of which determines the contract's nullity, that can be determined by the Courts according to Article 1421 of the Civil Code (legitimation to submit nullity's action). The void contract will not have any effect, but there will still be the possibility for the parties to convert it according to Article 1424 of the Civil Code.[257]

The previous considerations on the contract's voidance are valid also in case of lack of one or more of the essential requirements foreseen by Article 62(1).

[257] Article 1424 c.c. Conversione del contratto nullo: '*il contratto nullo può produrre gli effetti di un contratto diverso del quale contenga i requisiti di sostanza e di forma, qualora, avuto riguardo allo scopo perseguito dalle parti, debba ritenersi che esse lo avrebbero voluto se avessero conosciuto la nullità*'.

The Italian legislator establishes the principles the contracts will have to comply with: 'transparency, fair dealing, proportionality and mutual correspondence of the performances, with reference to the goods supplied'[258]. In practice, this provision refers to the cases of the mentioned article 3-bis 'contractual terms on improper burdens for the suppliers', where the selling typologies determining an abuse of dominant position or anti-competitive practice were forbidden according to the Law 287/1990 (Italian Antitrust Law).

Finally, it is useful to recall Article 7 of Directive 2011/7/EU of the European Parliament and of the Council of 16 February 2011 on combating late payment in commercial transactions, according to which in determining whether a contractual term or a practice relating to the date or period of payment is grossly unfair to the creditor, all circumstances of the case shall be considered, including any gross discrepancy from good commercial practice, contrary to good faith and fair dealing.

12.3 Forbidden contractual terms – Article 62(2)

Paragraph 2 puts in practice the principles indicated in Paragraph 1, establishing forbidden contractual terms:

a. it is forbidden to impose directly or indirectly purchase, selling, or other unjustifiably exacting contractual terms, as well as extra-contractual and retroactive terms.
   - These kind of forbidden terms are unfair commercial practices imposed by the distributors to the supplier, like for instance:
   - reductions, bonus, contributions, charge-off or compensation of commercial cooperation agreements to be paid in advance and with retroactive effect;
   - contributions to be paid by the supplier for the marketing expenses for the company's sign incurred by the distributor and/or contributions to be paid by the supplier for software compliance expenses of the distributor and/or accounting web standardisation (expenses for web invoicing);
   - contributions to be paid by the supplier for the lower profits deriving from reselling of his products;
   - payment of a flat rate by the supplier for the keeping in warehouse or other logistic activities, or for the inclusion of the supplier's products in the distributor's list price, or for the maintenance of the point of sale's assortment;
   - reductions, bonus, contributions only for complying with the terms of payment respect, provided by the supply contract;
   - possibility for the distributors to refuse and/or return the products or deduct penalties established by the contract without previous complaint to the supplier in the terms provided by the supply contract or, in absence of the ones foreseen by Article 1495 of the Civil Code.

[258] In the original version: *'la trasparenza, la correttezza, la proporzionalità e la reciproca corrispettività delle prestazioni, con riferimento ai beni forniti'.*

b. it is forbidden to apply objectively different conditions for equal performances: discriminatory conditions applied by distributors to suppliers of foodstuff belonging to the same product category.
c. it is forbidden for the parties to condition the contract's conclusion and/or execution, as well as the continuity and regularity of the same commercial relationships to the execution of performances that, for their nature or according to the commercial usage, do not have any connection with the object of both. In this regard, the forbidden contractual terms will be the ones establishing one or more performances not directly linked to the foodstuffs' sale, like for instance in the relationships between supplier and distributor, the request made by the supplier to the Large Distribution to incur the company's distribution costs; or, contrary, the imposition made by the Large Distribution to the supplier of the costs for the administrative services not directly linked to the product's sale.
d. it is forbidden to get undue one-sided performances, not justified by the nature or the content of the commercial relationships: for instance, the payment of a flat-rate by the supplier for the keeping in warehouse or other logistic activities, or for the inclusion of the supplier's products in the distributor's list price, or for the maintenance of the point of sale's assortment and for other referencing services;
e. it is forbidden to adopt any other unfair commercial dealing, taking into account all the commercial relationships that characterise the supply terms.

Concerning the last provision of letter e), the implementing Decree specifies in Article 4(1) that the definition of unfair commercial dealing includes the noncompliance with the principles of fair practices and unfair practices identified by the European Commission and the representatives on European level of the food supply chain within the High level Forum for a better functioning of the food supply chain, approved on the 29th of November 2011, document attached to the implementing Decree.

Paragraph 2 of Article 4 of the implementing Decree clarifies which other attitudes are included in the scope of Article 62(2). In particular, any attitude of the party who, taking advantage from its own bigger commercial power, imposes unjustifiably exacting contractual terms, is forbidden, included those terms that:
a. provide for one party the inclusion of services and/or incidental services with respect to the main object of the supply, even in case these are provided by third subjects, without any objective, direct and logic connection with the sale of the contract's product;
b. exclude the application of interests for late payment to the detriment of the creditor or exclude the compensation of expenses for the credit's recovery;
c. determine, in contrast to good faith and fairness principle, prices clearly under the production costs of the products object of the commercial transactions.

Another case of unfair commercial practice, according to Article 4(3) of the implementing Decree, is to insert a clause in the contract compulsorily imposing on the seller, after the delivery of the products, a minimum term before the issue of the invoice, except for the case of delivery of products in different parts during the same month, in which case the invoice can be issued only after the last delivery of the month.

Paragraph 4 of Article 3 provides that the unfairness of a commercial practice cannot be excluded for the only circumstance that a party is not able to prove that the other party abuses his/her own market or negotiation power in order to obtain an economic advantage not justified or unjustifiably exacting, resulting a detected diffusion of the practice one of the proof elements of the case.

It is important to underline that these forbidden terms are valid also for the contracts between foodstuff suppliers and Central Buying Offices representing Large Distribution chains, in accordance with the implementing Decree definitions that expressly includes this kind of contracts in the scope of Article 62 (Article 2(1)(l)).

The Central Buying Office is a structure organised by distribution companies operating in the same sector (foodstuffs, etc.) with the common target to obtain from the production companies raw materials, consumer goods or services necessary for the business activity. The Central Buying Office has therefore the main function of getting goods and services from the companies on behalf of its associates, a role that can be classified within Article 1731 of the Civil Code.[259] Besides the specific role of this structure, the Central Buying Office, on the basis of the agreements made with the producers and distributors, will be able to supply other kinds of services both to the associates and production companies, within the organising, administrative and commercial activity. On this subject, the Italian Finance Ministry Resolution no. 36/E mentioned above takes into consideration the operations between Central Buying Office and supply companies with costs to be paid by the latter:

a. business and marketing:
   - coordination and/or central management of the launch of products;
   - central management of the lists;
   - central management of the promotional operations' schedule on national level;
   - execution check of the agreed activities;
   - intervention on unfulfilling mandators;
   - implementation and check of activities defined in the framework agreement and others (check and sensitisation on lists' application, reaching invoiced target).

[259] Commission contract: Article 1731 c.c. Della commissione:'*Il contratto di commissione è un mandato che ha per oggetto l'acquisto o la vendita di beni per conto del committente e in nome del commissionario*'.

b. organisation activity:
   - organisation and coordination of commercial commissions for meetings with the mandator companies;
   - arrangement of offices with telephone, fax, photocopier, meeting room;
   - arrangement of documentation and any other business.
c. administration activity:
   - drafting of framework agreement;
   - disclosure to every company of the activities defined in the framework agreement;
   - disclosure of other information on other agreements and issues;
   - computation of the purchase data of every mandatory on every supplier;
   - check of the purchase data of every mandatory on every supplier;
   - collecting data and elaboration of commercial statistics.

12.4 Payment schedules – Article 62(3)

Paragraph 3 foresees that the payment of the foodstuffs' sale price has to be made within the time limit of 30 days for perishable food products and 60 days for all the other food products. In both cases, the term runs from the last day of the month in which the invoice has been received and the interests for late payment run automatically from the day subsequent to the term's deadline.

With the new provision, the legislator has therefore reduced the payment schedules for the contracts concerning perishable food products with respect to Article 4(3) of Decree 231/2002, that established a time limit of 60 days for these products, starting from the delivery or the collection of the products. As a consequence, Paragraphs 3 and 4 of Article 4 of Decree 231/2002 are expressly annulled by Paragraph 11 of Article 62.

According to the implementing Decree, the date from which the term runs will be identified with the date of receipt of the invoice, which is considered effectively certified only in case of hand delivery of the invoice, by sending the invoice with a registered letter with advice of delivery, with certified e-mail ('PEC') or with EDI system (Electronic Data Interchange), or any other equivalent system, as foreseen by tax law (Article 5(3)). When the date of receipt of the invoice is uncertain, it is considered the invoice is received on the date of delivery of the products.

It is important to underline that if the price is not paid with respect to the payment schedules, this circumstance should constitute an enforceable title to obtain a cease-and-desist order with reference to Articles 633 et seq. of the Italian Civil

Procedure Code. This interpretation is confirmed by the provision concerning the recovery procedures of unchallenged claims of Article 10 of Directive 2011/7/EU.[260]

In cases in which the contract does not establish any payment time limits, we deem that it is possible to apply Article 4(2) of Decree 231/2002, also confirmed by Directive 2011/7/EU, according to which the creditor is entitled to interest for late payment when the following time limits are expired:

- 30 days following the date of receipt by the debtor of the invoice or an equivalent request for payment;
- where the date of receipt of the invoice or the equivalent request for payment is uncertain, 30 days after the date of the receipt of the goods;
- where the debtor receives the invoice or the equivalent request for payment earlier than the goods or the services, 30 days after the date of the receipt of the goods or services.
- 30 days after the date of acceptance or verification, eventually provided for by statute or in the contract by which the conformity of the goods or services with the contract is to be ascertained, if the debtor receives the invoice or the equivalent request for payment earlier or on the date on which such acceptance or verification takes place.

Concerning the interests, they run automatically from the day subsequent to the payment term. The definitions of 'interest', 'statutory interest for late payment' and 'reference rate' are provided by the implementing Decree, Article 2(1)(f), (g) and (h):

f. *interests*: statutory interest for late payment or interest at a rate agreed upon between undertakings;
g. *statutory interest for late payment*: simple interest for late payment at a rate which is equal to the reference rate as defined by the in force national legislation of implementation of the European directives on combating late payment in commercial transactions;
h. *reference rate*: the interest rate as defined by the national legislation implementing the European directives on combating late payment in commercial transactions, applicable as follows:

[260] Article 10 Directive 2011/7/EU- Recovery procedures for unchallenged claims: '1. Member States shall ensure that an enforceable title can be obtained, including through an expedited procedure and irrespective of the amount of the debt, normally within 90 calendar days of the lodging of the creditor's action or application at the court or other competent authority, provided that the debt or aspects of the procedure are not disputed. Member States shall carry out this duty in accordance with their respective national laws, regulations and administrative provisions. 2. National laws, regulations and administrative provisions shall apply the same conditions for all creditors who are established in the Union. 3. When calculating the period referred to in Paragraph 1, the following shall not be taken into account: (a) periods for service of documents; (b) any delays caused by the creditor, such as periods devoted to correcting applications'.

- for the first half of the year in question the reference rate is the one in force on 1st January of that year;
- for the second half of the year in question the reference rate is the one in force on 1st July of that year.

Article 6 Paragraph 1 of the implementing Decree provides that the interests run from the day subsequent to the payment term, when the following conditions are respected:

- the creditor has fulfilled contractual and legal obligations;
- the creditor has not received within the term the amount due and the delay is due to the debtor.

Paragraph 3 of the same provision says that the interests are calculated by using the statutory interests for late payment or the reference rate agreed upon between undertakings, as long as that rate is not unfair for the creditor, taking into account the rises foreseen by the law. According to Paragraph 4 of Article 6, the statutory interests for late payment are calculated by using the reference rate indicated by the national legislation of implementation of the European directives on combating late payment in commercial transactions.

On this issue, we deem that the applicable rate of interest, that cannot be derogated by the parties, is the one provided by Article 5 of Decree 231/2002 mentioned above, that is to say the interest rate applied by the European Central Bank increased by 2 percentage points.[261]

12.5 Perishable foodstuffs – Article 62(4)

Paragraph 4 defines perishable foodstuffs:

a. agricultural, fish and pre-packaged food products with an expiring date and a lowest conservation time limit not longer than 60 days;
b. agricultural, fish and unpacked food products, including vegetables, even protected by a casing or cooled, not subjected to treatments aimed to lengthen the products' durability for a period longer than 60 days;
c. meat food products with particular physic-chemical features;
d. all kind of milk.

[261] Article 5 D. lgs. 231/2002: '*1. Salvo diverso accordo tra le parti, il saggio degli interessi, ai fini del presente decreto, è determinato in misura pari al saggio d'interesse del principale strumento di rifinanziamento della Banca centrale europea applicato alla sua più recente operazione di rifinanziamento principale effettuata il primo giorno di calendario del semestre in questione, maggiorato di sette punti percentuali. Il saggio di riferimento in vigore il primo giorno lavorativo della Banca centrale europea del semestre in questione si applica per i successivi sei mesi.*
2. Il Ministero dell'economia e delle finanze dà notizia del saggio di cui al comma 1, al netto della maggiorazione ivi prevista, curandone la pubblicazione nella Gazzetta Ufficiale della Repubblica italiana nel quinto giorno lavorativo di ciascun semestre solare'.

These categories of product have been already provided by Article 4 of Decree 231/2002 and by the Decree of the Minister of Productive Activities of 13 May 2003. All these provisions have been annulled by Paragraph 11 of Article 62.

12.6 Sanctions – Article 62(5), (6) and (7)

Paragraphs 5, 6 and 7 provide the fines in case of infringement of the provisions contained in Paragraphs 1, 2 and 3 of Article 62, except for the cases in which the infringement constitutes a criminal offence. We recall that a fine is the sanction provided by law for the infringement of a provision, identifying an administrative offence.

Article 8 of the Law 689/1981, modifying the criminal law system, regulates the administrative sanctions system applicable in case of more infringements of provisions. When several provisions, providing sanctions, are infringed by only one action or omission or when one provision is infringed several times, the sanction provided for the greatest infringement, will be applied increased until three times (so-called '*cumulo giuridico*'), according to Article 8(1) of L. 689/81. When the infringements are caused by more actions or omissions, the sanctions applied will be as many as the infringements.

In particular, the party who infringes the obligations provided by Paragraph 1 and does not conclude the contract using the mandatory written form, or in that contract one or more of the essential elements are missing, is subject to an administrative fine between € 516.00 and € 20,000.00. In such cases, the extent of the fine will be determined with reference to the goods' value.

The violation of the prohibitions imposed by Paragraph 2 for the fair construction of commercial relationships founded on principles of transparency, fair dealing, proportionality and mutual correspondence of the performances, expressed in Paragraph 1, is punished with an administrative fine between € 516.00 and € 3,000.00. In such cases, the sanction will be determined with reference to the benefit received by the party who has not respected the prohibitions imposed by Paragraph 2. For example, the economic benefit that the Large Distribution could obtain from the use of confidential information received by the supplier/producer of food products.

Finally, the failure to comply with the payment terms by the debtor, provided by Paragraph 3, is punished with an administrative fine between € 500.00 and € 500,000.00. In such cases, the extent of the fine is determined in accordance to the company's turnover, the recurrence and the delays' extent. The intent of the provision seems to be the protection of the creditor producers that are small and medium-sized enterprises (SMEs), because the extent of the sanction for the debtor (Large Distribution) is determined with respect to its turnover, that in general is quite high. This trend is confirmed by Directive 2011/7/EU, which in recital no. 6 emphasises that access to finance for small and medium-sized enterprises should be facilitated and that a legal and business behaviour supportive of timely payments in commercial transactions should be developed.[262]

12.7 Monitoring activity and imposition of sanctions – Article 62(8)

With Paragraphs 5 and 8, the legislator gave the Italian Antitrust Authority the task both to control the application of the entire regulation on commercial relationships concerning foodstuffs and to impose sanctions. The Antitrust Authority will provide automatically or on warnings made by any person interested to ascertain the infringements of the provisions of Paragraph 1 (no written form and/or no essential requirements and check of the contracts' nullity), Paragraph 2 (unfair dealing) and Paragraph 3 (late payment). In its investigation activity, the Antitrust Authority is allowed by law to use the Guardia di Finanza, the military corps dealing with customs, excise and tax crime, that will operate as judiciary police.

The Antitrust competence on this subject was already established by Article 3bis of the draft regulation 'Bersani ter' mentioned above. We deem that the reference to Law 689/1981 made by the legislator in this Paragraph is limited only to the Antitrust Authority power of imposing administrative sanctions.

In accordance with this new regulation and with what it is established in Article 12 et seq. of Law no. 287 of 1990 concerning 'Antitrust Authority's power on anti-competitive agreements and abuse of dominant position', the Antitrust Authority should provide an investigation procedure with a regulation in order to guarantee the debate between parties, the full knowledge of the acts and the verbalisation. This regulation will have to indicate interpretative guidelines to define the unfair dealing in violation of the principles contained in Paragraph 1 and the forbidden cases of Paragraph 2.

12.8 Reparation and injunction – Article 62(1)0

First of all, it is important to underline that Paragraph 10 gives the civil court both the competence for individual claims for damages caused by the infringements to

[262] Directive 2011/7/EU – recital no. 6: 'In its Communication of 25 June 2008 entitled "Think Small First – A Small Business Act for Europe", the Commission emphasised that small and medium-sized enterprises' (SMEs) access to finance should be facilitated and that a legal and business environment supportive of timely payments in commercial transactions should be developed. It should be noted that public authorities have a special responsibility in this regard. The criteria for the definition of SMEs are set out in Commission Recommendation 2003/361/EC of 6 May 2003 concerning the definition of micro, small and medium-sized enterprises'.

Article 62, to be activated in the ordinary judgments, and the claims for damages proposed by professional associations through class action.

The collective subjects that can introduce a class action for damages with reference to Article 62 infringements, are both the professional associations that are part of the National Economic and Labour Council (CNEL) or any other association that is representative on a national level, and the consumers associations that are part of the National Consumers Council (CNCU). In this regard, we do not understand the possibility for the consumers association to introduce such a claim for damages, since contracts with consumers are not included in the provision's scope, as provided by Paragraph 1.

The legitimation of the professional associations protecting collective interests was already provided by Article 8 of Decree 231/2002, where they could ask the competent judge:

> to ascertain the gross unfairness of the general conditions concerning the date for payment or the consequences of the delay and to forbid their practice; to adopt measures suitable to correct or remove the damages caused by the ascertained infringements; to order the publication of the decision in one or more national or local newspapers whenever the advertising could contribute to correct or remove the ascertained infringements' effects.

The claim for damages and the professional associations' legitimation are also considered in Article 7(1) and (5) of Directive 2011/7/EU on late payment. Paragraph 5 foresees that, in order to stop the common use of grossly unfair contractual terms or practices, the professional organisations representing companies or holding a lawful interest to represent companies, could take legal action before jurisdictional authorities or the competent administrative bodies.

The last part of Paragraph 10 gives the professional associations the possibility to take also emergency appeals according to Articles 699-bis et seq. of the Italian Civil Procedure Code to ask to stop the behaviours in breach of Paragraph 1 and 2 of Article 62, as it was provided by Paragraph 2 of Article 8 of Decree 231/2002.[263]

12.9 Abrogation – Article 62(11)

As already underlined in point 4 'Payment schedules' and in point 5 'Perishable foodstuffs' the provisions of Decree 231/2002 and the Decree of the Minister of Productive Activities of 13 May 2003 on the same issues have been abrogated.

[263] Article 8, Paragraph 2 of Decree 231/2002 *Tutela degli interessi collettivi: 'l'inibitoria è concessa, quando ricorrono giusti motivi di urgenza, ai sensi degli artt. 669 bis e segg. del Codice di procedura Civile'.*

12.10 Effectiveness and implementing decree – Article 62(11-bis)

In converting the decree in law, the Parliament has fixed the coming into force of Article 62 to 24th October 2012. This postponement has been planned in order to implement the new provisions with the inter-ministerial decree of the Minister of Agriculture together with the Minister of Economic Development, to be issued before the 25th of June 2012, as we already mentioned in the analysis of Article 62.

Annex

I

(Legislative acts)

REGULATIONS

REGULATION (EU) No 1151/2012 OF THE EUROPEAN PARLIAMENT AND OF THE COUNCIL

of 21 November 2012

on quality schemes for agricultural products and foodstuffs

THE EUROPEAN PARLIAMENT AND THE COUNCIL OF THE EUROPEAN UNION,

Having regard to the Treaty on the Functioning of the European Union, and in particular Article 43(2) and the first paragraph of Article 118 thereof,

Having regard to the proposal from the European Commission,

After transmission of the draft legislative act to the national parliaments,

Having regard to the opinion of the European Economic and Social Committee [1],

Having regard to the opinion of the Committee of the Regions [2],

Acting in accordance with the ordinary legislative procedure [3],

Whereas:

(1) The quality and diversity of the Union's agricultural, fisheries and aquaculture production is one of its important strengths, giving a competitive advantage to the Union's producers and making a major contribution to its living cultural and gastronomic heritage. This is due to the skills and determination of Union farmers and producers who have kept traditions alive while taking into account the developments of new production methods and material.

(2) Citizens and consumers in the Union increasingly demand quality as well as traditional products. They are also concerned to maintain the diversity of the agricultural production in the Union. This generates a demand for agricultural products or foodstuffs with identifiable specific characteristics, in particular those linked to their geographical origin.

(3) Producers can only continue to produce a diverse range of quality products if they are rewarded fairly for their effort. This requires that they are able to communicate to buyers and consumers the characteristics of their product under conditions of fair competition. It also requires them to be able to correctly identify their products on the marketplace.

(4) Operating quality schemes for producers which reward them for their efforts to produce a diverse range of quality products can benefit the rural economy. This is particularly the case in less favoured areas, in mountain areas and in the most remote regions, where the farming sector accounts for a significant part of the economy and production costs are high. In this way quality schemes are able to contribute to and complement rural development policy as well as market and income support policies of the common agricultural policy (CAP). In particular, they may contribute to areas in which the farming sector is of greater economic importance and, especially, to disadvantaged areas.

(5) The Europe 2020 policy priorities as set out in the Commission Communication entitled 'Europe 2020: A strategy for smart, sustainable and inclusive growth', include the aims of achieving a competitive economy based on knowledge and innovation and fostering a high-employment economy delivering social and territorial cohesion. Agricultural product quality policy should therefore provide producers with the right tools to better identify and promote those of their products that have specific characteristics while protecting those producers against unfair practices.

(1) OJ C 218, 23.7.2011, p. 114.
(2) OJ C 192, 1.7.2011, p. 28.
(3) Position of the European Parliament of 13 September 2012 (not yet published in the Official Journal) and decision of the Council of 13 November 2012.

(6) The set of complementary measures envisaged should respect the principles of subsidiarity and proportionality.

(7) Agricultural product quality policy measures are laid down in Council Regulation (EEC) No 1601/91 of 10 June 1991 laying down general rules on the definition, description and presentation of aromatized wines, aromatized wine-based drinks and aromatized wine-product cocktails [1]; Council Directive 2001/110/EC of 20 December 2001 relating to honey [2] and in particular in Article 2 thereof, Council Regulation (EC) No 247/2006 of 30 January 2006 laying down specific measures for agriculture in the outermost regions of the Union [3] and in particular in Article 14 thereof; Council Regulation (EC) No 509/2006 of 20 March 2006 on agricultural products and foodstuffs as traditional specialities guaranteed [4]; Council Regulation (EC) No 510/2006 of 20 March 2006 on the protection of geographical indications and designations of origin for agricultural products and foodstuffs [5]; Council Regulation (EC) No 1234/2007 of 22 October 2007 establishing a common organisation of agricultural markets and on specific provisions for certain agricultural products (Single CMO Regulation) [6] and in particular in Part II, Title II, Chapter I, Section I and in Section Ia, Subsection I thereof; Council Regulation (EC) No 834/2007 of 28 June 2007 on organic production and labelling of organic products [7]; and Regulation (EC) No 110/2008 of the European Parliament and of the Council of 15 January 2008 on the definition, description, presentation, labelling and the protection of geographical indications of spirit drinks [8].

(8) The labelling of agricultural products and foodstuffs should be subject to the general rules laid down in Directive 2000/13/EC of the European Parliament and of the Council of 20 March 2000 on the approximation of the laws of the Member States relating to the labelling, presentation and advertising of foodstuffs [9], and in particular the provisions aimed at preventing labelling that may confuse or mislead consumers.

(9) The Communication from the Commission to the European Parliament, the Council, the European Economic and Social Committee and the Committee of the Regions on agricultural product quality policy identified the achievement of a greater overall coherence and consistency of agricultural product quality policy as a priority.

(10) The geographical indications scheme for agricultural products and foodstuffs and the traditional specialities guaranteed scheme have certain common objectives and provisions.

(11) The Union has for some time been pursuing an approach that aims to simplify the regulatory framework of the CAP. This approach should also be applied to regulations in the field of agricultural product quality policy, without, in so doing, calling into question the specific characteristics of those products.

(12) Some regulations that form part of the agricultural product quality policy have been reviewed recently but are not yet fully implemented. As a result, they should not be included in this Regulation. However, they may be incorporated at a later stage, once the legislation has been fully implemented.

(13) In the light of the aforementioned considerations, the following provisions should be amalgamated into a single legal framework comprising the new or updated provisions of Regulations (EC) No 509/2006 and (EC) No 510/2006 and those provisions of Regulations (EC) No 509/2006 and (EC) No 510/2006 that are maintained.

(14) In the interests of clarity and transparency, Regulations (EC) No 509/2006 and (EC) No 510/2006 should therefore be repealed and replaced by this Regulation.

(15) The scope of this Regulation should be limited to the agricultural products intended for human consumption listed in Annex I to the Treaty and to a list of products outside the scope of that Annex that are closely linked to agricultural production or to the rural economy.

(16) The rules provided for in this Regulation should apply without affecting existing Union legislation on wines, aromatised wines, spirit drinks, product of organic farming, or outermost regions.

(17) The scope for designations of origin and geographical indications should be limited to products for which an intrinsic link exists between product or foodstuff characteristics and geographical origin. The inclusion in the current scheme of only certain types of chocolate as confectionery products is an anomaly that should be corrected.

(18) The specific objectives of protecting designations of origin and geographical indications are securing a fair return for farmers and producers for the qualities and characteristics of a given product, or of its mode of production, and providing clear information on products with specific characteristics linked to geographical origin, thereby enabling consumers to make more informed purchasing choices.

[1] OJ L 149, 14.6.1991, p. 1.
[2] OJ L 10, 12.1.2002, p. 47.
[3] OJ L 42, 14.2.2006, p. 1.
[4] OJ L 93, 31.3.2006, p. 1.
[5] OJ L 93, 31.3.2006, p. 12.
[6] OJ L 299, 16.11.2007, p. 1.
[7] OJ L 189, 20.7.2007, p. 1.
[8] OJ L 39, 13.2.2008, p. 16.
[9] OJ L 109, 6.5.2000, p. 29.

(19) Ensuring uniform respect throughout the Union for the intellectual property rights related to names protected in the Union is a priority that can be achieved more effectively at Union level.

(20) A Union framework that protects designations of origin and geographical indications by providing for their inclusion on a register facilitates the development of those instruments, since the resulting, more uniform, approach ensures fair competition between the producers of products bearing such indications and enhances the credibility of the products in the consumers' eyes. Provision should be made for the development of designations of origin and geographical indications at Union level and for promoting the creation of mechanisms for their protection in third countries in the framework of the World Trade Organisation (WTO) or multilateral and bilateral agreements, thereby contributing to the recognition of the quality of products and of their model of production as a factor that adds value.

(21) In the light of the experience gained from the implementation of Council Regulation (EEC) No 2081/92 of 14 July 1992 on the protection of geographical indications and designations of origin for agricultural products and foodstuffs [1] and Regulation (EC) No 510/2006, there is a need to address certain issues, to clarify and simplify some rules and to streamline the procedures of this scheme.

(22) In the light of existing practice, the two different instruments for identifying the link between the product and its geographical origin, namely the protected designation of origin and the protected geographical indication, should be further defined and maintained. Without changing the concept of those instruments, some modifications to the definitions should be adopted in order to better take into account the definition of geographical indications laid down in the Agreement on Trade-Related Aspects of Intellectual Property Rights and to make them simpler and clearer for operators to understand.

(23) An agricultural product or foodstuff bearing such a geographical description should meet certain conditions set out in a specification, such as specific requirements aimed at protecting the natural resources or landscape of the production area or improving the welfare of farm animals.

(24) To qualify for protection in the territories of Member States, designations of origin and geographical indications should be registered only at Union level. With effect from the date of application for such registration at Union level, Member States should be able to grant transitional protection at national level without affecting intra-Union or international trade. The protection afforded by this Regulation upon registration, should be equally available to designations of origin and geographical indications of third countries that meet the corresponding criteria and that are protected in their country of origin.

(25) The registration procedure at Union level should enable any natural or legal person with a legitimate interest from a Member State, other than the Member State of the application, or from a third country, to exercise their rights by notifying their opposition.

(26) Entry in the register of protected designations of origin and protected geographical indications should also provide information to consumers and to those involved in trade.

(27) The Union negotiates international agreements, including those concerning the protection of designations of origin and geographical indications, with its trade partners. In order to facilitate the provision to the public of information about the names so protected, and in particular to ensure protection and control of the use to which those names are put, the names may be entered in the register of protected designations of origin and protected geographical indications. Unless specifically identified as designations of origin in such international agreements, the names should be entered in the register as protected geographical indications.

(28) In view of their specific nature, special provisions concerning labelling should be adopted in respect of protected designations of origin and protected geographical indications that require producers to use the appropriate Union symbols or indications on packaging. In the case of Union names, the use of such symbols or indications should be made obligatory in order to make this category of products, and the guarantees attached to them, better known to consumers and in order to permit easier identification of these products on the market, thereby facilitating checks. Taking into account the requirements of the WTO, the use of such symbols or indications should be made voluntary for third-country geographical indications and designations of origin.

(29) Protection should be granted to names included in the register with the aim of ensuring that they are used fairly and in order to prevent practices liable to mislead consumers. In addition, the means of ensuring that geographical indications and designations of origin are protected should be clarified, particularly as regards the role of producer groups and competent authorities of Member States.

(30) Provision should be made for specific derogations that permit, for transitional periods, the use of a registered name alongside other names. Those derogations should

[1] OJ L 208, 24.7.1992, p. 1.

be simplified and clarified. In certain cases, in order to overcome temporary difficulties and with the long-term objective of ensuring that all producers comply with the specifications, those derogations may be granted for a period of up to 10 years.

(31) The scope of the protection granted under this Regulation should be clarified, in particular with regard to those limitations on registration of new trade marks set out in Directive 2008/95/EC of the European Parliament and of the Council of 22 October 2008 to approximate the laws of the Member States relating to trade marks ([1]) that conflict with the registration of protected designations of origin and protected geographical indications as is already the case for the registration of new trade marks at Union level. Such clarification is also necessary with regard to the holders of prior rights in intellectual property, in particular those concerning trade marks and homonymous names registered as protected designations of origin or as protected geographical indications.

(32) Protection of designations of origin and geographical indications should be extended to the misuse, imitation and evocation of the registered names on goods as well as on services in order to ensure a high level of protection and to align that protection with that which applies to the wine sector. When protected designations of origin or protected geographical indications are used as ingredients, the Commission Communication entitled 'Guidelines on the labelling of foodstuffs using protected designations of origin (PDOs) or protected geographical indications (PGIs) as ingredients' should be taken into account.

(33) The names already registered under Regulation (EC) No 510/2006 on 3 January 2013 should continue to be protected under this Regulation and they should be automatically included in the register.

(34) The specific objective of the scheme for traditional specialities guaranteed is to help the producers of traditional products to communicate to consumers the value-adding attributes of their product. However, as only a few names have been registered, the current scheme for traditional specialities guaranteed has failed to realise its potential. Current provisions should therefore be improved, clarified and sharpened in order to make the scheme more understandable, operational and attractive to potential applicants.

(35) The current scheme provides the option to register a name for identification purposes without reservation of the name in the Union. As this option has not been well understood by stakeholders and since the function of identifying traditional products can be better achieved at Member State or regional level in application of the principle of subsidiarity, this option should be discontinued. In the light of experience, the scheme should only deal with the reservation of names across the Union.

(36) To ensure that names of genuine traditional products are registered under the scheme, the criteria and conditions for registration of a name should be adapted, in particular those concerning the definition of 'traditional', which should cover products that have been produced for a significant period of time.

(37) To ensure that traditional specialities guaranteed comply with their specification and are consistent, producers organised into groups should themselves define the product in a specification. The option of registering a name as a traditional speciality guaranteed should be open to third-country producers.

(38) To qualify for reservation, traditional specialities guaranteed should be registered at Union level. The entry in the register should also provide information to consumers and to those involved in the trade.

(39) In order to avoid creating unfair conditions of competition, any producer, including a third-country producer, should be able to use a registered name of a traditional speciality guaranteed, provided that the product concerned complies with the requirements of the relevant specification and the producer is covered by a system of controls. For traditional specialities guaranteed produced within the Union, the Union symbol should be indicated on the labelling and it should be possible to associate it with the indication 'traditional speciality guaranteed'.

(40) In order to protect registered names from misuse, or from practices that might mislead consumers, their use should be reserved.

(41) For those names already registered under Regulation (EC) No 509/2006 that, on 3 January 2013, would otherwise not be covered by the scope of this Regulation, the terms of use laid down in Regulation (EC) No 509/2006 should continue to apply for a transitional period.

(42) A procedure should be introduced for registering names that are registered without reservation of name pursuant to Regulation (EC) No 509/2006, enabling them to be registered with reservation of name.

(43) Provision should also be made for transitional measures applicable to registration applications received by the Commission before 3 January 2013.

(44) A second tier of quality systems, based on quality terms which add value, which can be communicated on the internal market and which are to be applied voluntarily,

([1]) OJ L 299, 8.11.2008, p. 25.

should be introduced. Those optional quality terms should refer to specific horizontal characteristics, with regard to one or more categories of products, farming methods or processing attributes which apply in specific areas. The optional quality term 'mountain product' has met the conditions up to now and will add value to the product on the market. In order to facilitate the application of Directive 2000/13/EC where the labelling of foodstuffs may give rise to consumer confusion in relation to optional quality terms, including in particular 'mountain products', the Commission may adopt guidelines.

(45) In order to provide mountain producers with an effective tool to better market their product and to reduce the actual risks of consumer confusion as to the mountain provenance of products in the market place, provision should be made for the definition at Union level of an optional quality term for mountain products. The definition of mountain areas should build on the general classification criteria employed to identify a mountain area in Council Regulation (EC) No 1257/1999 of 17 May 1999 on support for rural development from the European Agricultural Guidance and Guarantee Fund (EAGGF) (1).

(46) The added value of the geographical indications and traditional specialities guaranteed is based on consumer trust. It is only credible if accompanied by effective verification and controls. Those quality schemes should be subject to a monitoring system of official controls, in line with the principles set out in Regulation (EC) No 882/2004 of the European Parliament and of the Council of 29 April 2004 on official controls performed to ensure the verification of compliance with feed and food law, animal health and animal welfare rules (2), and should include a system of checks at all stages of production, processing and distribution. In order to help Member States to better apply provisions of Regulation (EC) No 882/2004 for the controls of geographical indications and traditional specialities guaranteed, references to the most relevant articles should be mentioned in this Regulation.

(47) To guarantee to the consumer the specific characteristics of geographical indications and traditional specialities guaranteed, operators should be subject to a system that verifies compliance with the product specification.

(48) In order to ensure that they are impartial and effective, the competent authorities should meet a number of operational criteria. Provisions on delegating some competences of performing specific control tasks to control bodies should be envisaged.

(49) European standards (EN standards) developed by the European Committee for Standardisation (CEN) and international standards developed by the International Organisation for Standardisation (ISO) should be used for the accreditation of the control bodies as well as by those bodies for their operations. The accreditation of those bodies should take place in accordance with Regulation (EC) No 765/2008 of the European Parliament and of the Council of 9 July 2008 setting out the requirements for accreditation and market surveillance relating to the marketing of products (3).

(50) Information on control activities for geographical indications and traditional specialities guaranteed should be included in the multiannual national control plans and annual report prepared by the Member States in accordance with Regulation (EC) No 882/2004.

(51) Member States should be authorised to charge a fee to cover the costs incurred.

(52) Existing rules concerning the continued use of names that are generic should be clarified so that generic terms that are similar to or form part of a name or term that is protected or reserved should retain their generic status.

(53) The date for establishing the seniority of a trade mark and of a designation of origin or a geographical indication should be that of the date of application of the trade mark for registration in the Union or in the Member States and the date of application for protection of a designation of origin or a geographical indication to the Commission.

(54) The provisions dealing with the refusal or coexistence of a designation of origin or a geographical indication on the ground of conflict with a prior trade mark should continue to apply.

(55) The criteria by which subsequent trade marks should be refused or, if registered, invalidated on the ground that they conflict with a prior designation of origin or geographical indication should correspond to the scope of protection of designation of origin or a geographical indication laid down.

(56) The provisions of systems establishing intellectual property rights, and particularly of those established by the quality scheme for designations of origin and geographical indications or those established under trade mark law, should not be affected by the reservation of names and the establishment of indications and symbols pursuant to the quality schemes for traditional specialities guaranteed and for optional quality terms.

(1) OJ L 160, 26.6.1999, p. 80.
(2) OJ L 165, 30.4.2004, p. 1.
(3) OJ L 218, 13.8.2008, p. 30.

(57) The role of groups should be clarified and recognised. Groups play an essential role in the application process for the registration of names of designations of origin and geographical indications and traditional specialities guaranteed, as well as in the amendment of specifications and cancellation requests. The group can also develop activities related to the surveillance of the enforcement of the protection of the registered names, the compliance of the production with the product specification, the information and promotion of the registered name as well as, in general, any activity aimed at improving the value of the registered names and effectiveness of the quality schemes. Moreover, it should monitor the position of the products on the market. Nevertheless, these activities should not facilitate nor lead to anti-competitive conduct incompatible with Articles 101 and 102 of the Treaty.

(58) To ensure that registered names of designations of origin and geographical indications and traditional specialities guaranteed meet the conditions laid down by this Regulation, applications should be examined by the national authorities of the Member State concerned, in compliance with minimum common provisions, including a national opposition procedure. The Commission should subsequently scrutinise applications to ensure that there are no manifest errors and that Union law and the interests of stakeholders outside the Member State of application have been taken into account.

(59) Registration as designations of origin, geographical indications and traditional specialities guaranteed should be open to names that relate to products originating in third countries and that satisfy the conditions laid down by this Regulation.

(60) The symbols, indications and abbreviations identifying participation in a quality scheme, and the rights therein pertaining to the Union, should be protected in the Union as well as in third countries with the aim of ensuring that they are used on genuine products and that consumers are not misled as to the qualities of products. Furthermore, in order for the protection to be effective, the Commission should have recourse to reasonable budget resources on a centralised basis within the framework of Council Regulation (EC) No 1698/2005 of 20 September 2005 on support for rural development by the European Agricultural Fund for Rural Development (EAFRD) [1] and in accordance with Article 5 of Council Regulation (EC) No 1290/2005 of 21 June 2005 on the financing of the common agricultural policy [2].

(61) The registration procedure for protected designations of origin, protected geographical indications and traditional specialities guaranteed, including the scrutiny and the opposition periods, should be shortened and improved, in particular as regards decision making. The Commission, in certain circumstances acting with the assistance of Member States, should be responsible for decision-making on registration. Procedures should be laid down to allow the amendment of product specifications after registration and the cancellation of registered names, in particular if the product no longer complies with the corresponding product specification or if a name is no longer used in the market place.

(62) In order to facilitate cross-border applications for joint registration of protected designations of origin, protected geographical indications or traditional specialities guaranteed, provision should be made for appropriate procedures.

(63) In order to supplement or amend certain non-essential elements of this Regulation, the power to adopt acts in accordance with Article 290 of the Treaty should be delegated to the Commission in respect of supplementing the list of products set out in Annex I to this Regulation; establishing the restrictions and derogations with regard to the sourcing of feed in the case of a designation of origin; establishing restrictions and derogations with regard to the slaughtering of live animals or with regard to the sourcing of raw materials; laying down rules which limit the information contained in the product specification; establishing the Union symbols; laying down additional transitional rules in order to protect the rights and legitimate interests of producers or stakeholders concerned; laying down further details on the eligibility criteria for the names of traditional specialities guaranteed; laying down detailed rules relating to the criteria for optional quality terms; reserving an additional optional quality term, laying down its conditions of use and amending those conditions; laying down derogations to the use of the term 'mountain product' and establishing the methods of production, and other criteria relevant for the application of that optional quality term, in particular, laying down the conditions under which raw materials or feedstuffs are permitted to come from outside the mountain areas; laying down additional rules for determining the generic status of terms in the Union; laying down rules for determining the use of the name of a plant variety or of an animal breed; defining the rules for carrying out the national objection procedure for joint applications concerning more than one national territory; and for

[1] OJ L 277, 21.10.2005, p. 1.

[2] OJ L 209, 11.8.2005, p. 1.

complementing the rules of the application process, the opposition process, the amendment application process and the cancellation process in general. It is of particular importance that the Commission carry out appropriate consultations during its preparatory work, including at expert level. The Commission, when preparing and drawing up delegated acts, should ensure a simultaneous, timely and appropriate transmission of relevant documents to the European Parliament and to the Council.

(64) In order to ensure uniform conditions for the implementation of this Regulation, implementing powers should be conferred on the Commission as regards laying down rules on the form of the product specification; laying down detailed rules on the form and content of the register of protected designations of origin and protected geographical indications; defining the technical characteristics of the Union symbols and indications as well as the rules on their use on products, including the appropriate linguistic versions to be used; granting and extending transitional periods for temporary derogations for use of protected designations of origin and protected geographical indication; laying down detailed rules on the form and content of the register of traditional specialities guaranteed; laying down rules for the protection of traditional specialities guaranteed; laying down all measures relating to forms, procedures and other technical details for the application of Title IV; laying down rules for the use of optional quality terms; laying down rules for the uniform protection of indications, abbreviations and symbols referring to the quality schemes; laying down detailed rules on the procedure, form and presentation of applications for registration and of oppositions; rejecting the application; deciding on the registration of a name if an agreement has not been reached; laying down detailed rules on the procedure, form and presentation of an amendment application; cancelling the registration of a protected designation of origin, a protected geographical indication or a traditional speciality guaranteed; and laying down detailed rules on the procedure and form of the cancellation process and on the presentation of the requests for cancellation. Those powers should be exercised in accordance with Regulation (EU) No 182/2011 of the European Parliament and of the Council of 16 February 2011 laying down the rules and general principles concerning mechanisms for control by Member States of the Commission's exercise of implementing powers [1].

(65) In respect of establishing and maintaining registers of protected designations of origin, protected geographical indications and traditional specialties guaranteed, recognised under this scheme; defining the means by which the name and address of product certification bodies are to be made public; and registering a name if there is no notice of opposition or no admissible reasoned statement of opposition or in the case there is one the agreement has been reached, the Commission should be empowered to adopt implementing acts without applying Regulation (EU) No 182/2011,

[1] OJ L 55, 28.2.2011, p. 13.

HAVE ADOPTED THIS REGULATION:

TITLE I

GENERAL PROVISIONS

Article 1

Objectives

1. This Regulation aims to help producers of agricultural products and foodstuffs to communicate the product characteristics and farming attributes of those products and foodstuffs to buyers and consumers, thereby ensuring:

(a) fair competition for farmers and producers of agricultural products and foodstuffs having value-adding characteristics and attributes;

(b) the availability to consumers of reliable information pertaining to such products;

(c) respect for intellectual property rights; and

(d) the integrity of the internal market.

The measures set out in this Regulation are intended to support agricultural and processing activities and the farming systems associated with high quality products, thereby contributing to the achievement of rural development policy objectives.

2. This Regulation establishes quality schemes which provide the basis for the identification and, where appropriate, protection of names and terms that, in particular, indicate or describe agricultural products with:

(a) value-adding characteristics; or

(b) value-adding attributes as a result of the farming or processing methods used in their production, or of the place of their production or marketing.

Article 2

Scope

1. This Regulation covers agricultural products intended for human consumption listed in Annex I to the Treaty and other agricultural products and foodstuffs listed in Annex I to this Regulation.

In order to take into account international commitments or new production methods or material, the Commission shall be empowered to adopt delegated acts, in accordance with Article 56, supplementing the list of products set out in Annex I to this Regulation. Such products shall be closely linked to agricultural products or to the rural economy.

2. This Regulation shall not apply to spirit drinks, aromatised wines or grapevine products as defined in Annex XIb to Regulation (EC) No 1234/2007, with the exception of wine-vinegars.

3. This Regulation shall apply without prejudice to other specific Union provisions relating to the placing of products on the market and, in particular, to the single common organisation of the markets, and to food labelling.

4. Directive 98/34/EC of the European Parliament and of the Council of 22 June 1998 laying down a procedure for the provision of information in the field of technical standards and regulations and of rules on Information Society services (1) shall not apply to the quality schemes established by this Regulation.

Article 3

Definitions

For the purposes of this Regulation the following definitions shall apply:

(1) 'quality schemes' means the schemes established under Titles II, III and IV;

(2) 'group' means any association, irrespective of its legal form, mainly composed of producers or processors working with the same product;

(3) 'traditional' means proven usage on the domestic market for a period that allows transmission between generations; this period is to be at least 30 years;

(4) 'labelling' means any words, particulars, trade marks, brand name, pictorial matter or symbol relating to a foodstuff and placed on any packaging, document, notice, label, ring or collar accompanying or referring to such foodstuff;

(5) 'specific character' in relation to a product means the characteristic production attributes which distinguish a product clearly from other similar products of the same category;

(6) 'generic terms ' means the names of products which, although relating to the place, region or country where the product was originally produced or marketed, have become the common name of a product in the Union;

(7) 'production step' means production, processing or preparation;

(8) 'processed products' means foodstuffs resulting from the processing of unprocessed products. Processed products may contain ingredients that are necessary for their manufacture or to give them specific characteristics.

(1) OJ L 204, 21.7.1998, p. 37.

TITLE II

PROTECTED DESIGNATIONS OF ORIGIN AND PROTECTED GEOGRAPHICAL INDICATIONS

Article 4

Objective

A scheme for protected designations of origin and protected geographical indications is established in order to help producers of products linked to a geographical area by:

(a) securing fair returns for the qualities of their products;

(b) ensuring uniform protection of the names as an intellectual property right in the territory of the Union;

(c) providing clear information on the value-adding attributes of the product to consumers.

Article 5

Requirements for designations of origin and geographical indications

1. For the purpose of this Regulation, 'designation of origin' is a name which identifies a product:

(a) originating in a specific place, region or, in exceptional cases, a country;

(b) whose quality or characteristics are essentially or exclusively due to a particular geographical environment with its inherent natural and human factors; and

(c) the production steps of which all take place in the defined geographical area.

2. For the purpose of this Regulation, 'geographical indication' is a name which identifies a product:

(a) originating in a specific place, region or country;

(b) whose given quality, reputation or other characteristic is essentially attributable to its geographical origin; and

(c) at least one of the production steps of which take place in the defined geographical area.

3. Notwithstanding paragraph 1, certain names shall be treated as designations of origin even though the raw materials for the products concerned come from a geographical area larger than, or different from, the defined geographical area, provided that:

(a) the production area of the raw materials is defined;

(b) special conditions for the production of the raw materials exist;

(c) there are control arrangements to ensure that the conditions referred to in point (b) are adhered to; and

(d) the designations of origin in question were recognised as designations of origin in the country of origin before 1 May 2004.

Only live animals, meat and milk may be considered as raw materials for the purposes of this paragraph.

4. In order to take into account the specific character of production of products of animal origin, the Commission shall be empowered to adopt delegated acts, in accordance with Article 56, concerning restrictions and derogations with regard to the sourcing of feed in the case of a designation of origin.

In addition, in order to take into account the specific character of certain products or areas, the Commission shall be empowered to adopt delegated acts in accordance with Article 56, concerning restrictions and derogations with regard to the slaughtering of live animals or with regard to the sourcing of raw materials.

These restrictions and derogations shall, based on objective criteria, take into account quality or usage and recognised know-how or natural factors.

Article 6

Generic nature, conflicts with names of plant varieties and animal breeds, with homonyms and trade marks

1. Generic terms shall not be registered as protected designations of origin or protected geographical indications.

2. A name may not be registered as a designation of origin or geographical indication where it conflicts with a name of a plant variety or an animal breed and is likely to mislead the consumer as to the true origin of the product.

3. A name proposed for registration that is wholly or partially homonymous with a name already entered in the register established under Article 11 may not be registered unless there is sufficient distinction in practice between the conditions of local and traditional usage and presentation of the homonym registered subsequently and the name already entered in the register, taking into account the need to ensure equitable treatment of the producers concerned and that consumers are not misled.

A homonymous name which misleads the consumer into believing that products come from another territory shall not be registered even if the name is accurate as far as the actual territory, region or place of origin of the products in question is concerned.

4. A name proposed for registration as a designation of origin or geographical indication shall not be registered where, in the light of a trade mark's reputation and renown and the length of time it has been used, registration of the name proposed as the designation of origin or geographical indication would be liable to mislead the consumer as to the true identity of the product.

Article 7

Product specification

1. A protected designation of origin or a protected geographical indication shall comply with a specification which shall include at least:

(a) the name to be protected as a designation of origin or geographical indication, as it is used, whether in trade or in common language, and only in the languages which are or were historically used to describe the specific product in the defined geographical area;

(b) a description of the product, including the raw materials, if appropriate, as well as the principal physical, chemical, microbiological or organoleptic characteristics of the product;

(c) the definition of the geographical area delimited with regard to the link referred to in point (f)(i) or (ii) of this paragraph, and, where appropriate, details indicating compliance with the requirements of Article 5(3);

(d) evidence that the product originates in the defined geographical area referred to in Article 5(1) or (2);

(e) a description of the method of obtaining the product and, where appropriate, the authentic and unvarying local methods as well as information concerning packaging, if the applicant group so determines and gives sufficient product-specific justification as to why the packaging must take place in the defined geographical area to safeguard quality, to ensure the origin or to ensure control, taking into account Union law, in particular that on the free movement of goods and the free provision of services;

(f) details establishing the following:

(i) the link between the quality or characteristics of the product and the geographical environment referred to in Article 5(1); or

(ii) where appropriate, the link between a given quality, the reputation or other characteristic of the product and the geographical origin referred to in Article 5(2);

(g) the name and address of the authorities or, if available, the name and address of bodies verifying compliance with the provisions of the product specification pursuant to Article 37 and their specific tasks;

(h) any specific labelling rule for the product in question.

2. In order to ensure that product specifications provide relevant and succinct information, the Commission shall be empowered to adopt delegated acts, in accordance with Article 56, laying down rules which limit the information contained in the specification referred to in paragraph 1 of this Article, where such a limitation is necessary to avoid excessively voluminous applications for registration.

The Commission may adopt implementing acts laying down rules on the form of the specification. Those implementing acts shall be adopted in accordance with the examination procedure referred to in Article 57(2).

Article 8

Content of application for registration

1. An application for registration of a designation of origin or geographical indication pursuant to Article 49(2) or (5) shall include at least:

(a) the name and address of the applicant group and of the authorities or, if available, bodies verifying compliance with the provisions of the product specification;

(b) the product specification provided for in Article 7;

(c) a single document setting out the following:

(i) the main points of the product specification: the name, a description of the product, including, where appropriate, specific rules concerning packaging and labelling, and a concise definition of the geographical area;

(ii) a description of the link between the product and the geographical environment or geographical origin referred to in Article 5(1) or (2), as the case may be, including, where appropriate, the specific elements of the product description or production method justifying the link.

An application as referred to in Article 49(5) shall, in addition, include proof that the name of the product is protected in its country of origin.

2. An application dossier referred to in Article 49(4) shall comprise:

(a) the name and address of the applicant group;

(b) the single document referred to in point (c) of paragraph 1 of this Article;

(c) a declaration by the Member State that it considers that the application lodged by the applicant group and qualifying for the favourable decision meets the conditions of this Regulation and the provisions adopted pursuant thereto;

(d) the publication reference of the product specification.

Article 9

Transitional national protection

A Member State may, on a transitional basis only, grant protection to a name under this Regulation at national level, with effect from the date on which an application is lodged with the Commission.

Such national protection shall cease on the date on which either a decision on registration under this Regulation is taken or the application is withdrawn.

Where a name is not registered under this Regulation, the consequences of such national protection shall be the sole responsibility of the Member State concerned.

The measures taken by Member States under the first paragraph shall produce effects at national level only, and they shall have no effect on intra-Union or international trade.

Article 10

Grounds for opposition

1. A reasoned statement of opposition as referred to in Article 51(2) shall be admissible only if it is received by the Commission within the time limit set out in that paragraph and if it:

(a) shows that the conditions referred to in Article 5 and Article 7(1) are not complied with;

(b) shows that the registration of the name proposed would be contrary to Article 6(2), (3) or (4);

(c) shows that the registration of the name proposed would jeopardise the existence of an entirely or partly identical name or of a trade mark or the existence of products which have been legally on the market for at least five years preceding the date of the publication provided for in point (a) of Article 50(2); or

(d) gives details from which it can be concluded that the name for which registration is requested is a generic term.

2. The grounds for opposition shall be assessed in relation to the territory of the Union.

Article 11

Register of protected designations of origin and protected geographical indications

1. The Commission shall adopt implementing acts, without applying the procedure referred to in Article 57(2), establishing and maintaining a publicly accessible updated register of protected designations of origin and protected geographical indications recognised under this scheme.

2. Geographical indications pertaining to products of third countries that are protected in the Union under an international agreement to which the Union is a contracting party may be entered in the register. Unless specifically identified in the said agreement as protected designations of origin under this Regulation, such names shall be entered in the register as protected geographical indications.

3. The Commission may adopt implementing acts laying down detailed rules on the form and content of the register. Those implementing acts shall be adopted in accordance with the examination procedure referred to in Article 57(2).

4. The Commission shall make public and regularly update the list of the international agreements referred to in paragraph 2 as well as the list of geographical indications protected under those agreements.

Article 12

Names, symbols and indications

1. Protected designations of origin and protected geographical indications may be used by any operator marketing a product conforming to the corresponding specification.

2. Union symbols designed to publicise protected designations of origin and protected geographical indications shall be established.

3. In the case of products originating in the Union that are marketed under a protected designation of origin or a protected geographical indication registered in accordance with the procedures laid down in this Regulation, the Union symbols associated with them shall appear on the labelling. In addition, the registered name of the product should appear in the same field of vision. The indications 'protected designation of origin' or 'protected geographical indication' or the corresponding abbreviations 'PDO' or 'PGI' may appear on the labelling.

4. In addition, the following may also appear on the labelling: depictions of the geographical area of origin, as referred to in Article 5, and text, graphics or symbols referring to the Member State and/or region in which that geographical area of origin is located.

5. Without prejudice to Directive 2000/13/EC, the collective geographical marks referred to in Article 15 of Directive 2008/95/EC may be used on labels, together with the protected designation of origin or protected geographical indication.

6. In the case of products originating in third countries marketed under a name entered in the register, the indications referred to in paragraph 3 or the Union symbols associated with them may appear on the labelling.

7. In order to ensure that the appropriate information is communicated to the consumer, the Commission shall be empowered to adopt delegated acts, in accordance with Article 56, establishing the Union symbols.

The Commission may adopt implementing acts defining the technical characteristics of the Union symbols and indications as well as the rules of their use on the products marketed under a protected designation of origin or a protected geographical indication, including rules concerning the appropriate linguistic versions to be used. Those implementing acts shall be adopted in accordance with the examination procedure referred to in Article 57(2).

Article 13

Protection

1. Registered names shall be protected against:

(a) any direct or indirect commercial use of a registered name in respect of products not covered by the registration where those products are comparable to the products registered under that name or where using the name exploits the reputation of the protected name, including when those products are used as an ingredient;

(b) any misuse, imitation or evocation, even if the true origin of the products or services is indicated or if the protected name is translated or accompanied by an expression such as 'style', 'type', 'method', 'as produced in', 'imitation' or similar, including when those products are used as an ingredient;

(c) any other false or misleading indication as to the provenance, origin, nature or essential qualities of the product that is used on the inner or outer packaging, advertising material or documents relating to the product concerned, and the packing of the product in a container liable to convey a false impression as to its origin;

(d) any other practice liable to mislead the consumer as to the true origin of the product.

Where a protected designation of origin or a protected geographical indication contains within it the name of a product which is considered to be generic, the use of that generic name shall not be considered to be contrary to points (a) or (b) of the first subparagraph.

2. Protected designations of origin and protected geographical indications shall not become generic.

3. Member States shall take appropriate administrative and judicial steps to prevent or stop the unlawful use of protected designations of origin and protected geographical indications, as referred to in paragraph 1, that are produced or marketed in that Member State.

To that end Member States shall designate the authorities that are responsible for taking these steps in accordance with procedures determined by each individual Member State.

These authorities shall offer adequate guarantees of objectivity and impartiality, and shall have at their disposal the qualified staff and resources necessary to carry out their functions.

Article 14

Relations between trade marks, designations of origin and geographical indications

1. Where a designation of origin or a geographical indication is registered under this Regulation, the registration of a trade mark the use of which would contravene Article 13(1) and which relates to a product of the same type shall be refused if the application for registration of the trade mark is submitted after the date of submission of the registration application in respect of the designation of origin or the geographical indication to the Commission.

Trade marks registered in breach of the first subparagraph shall be invalidated.

The provisions of this paragraph shall apply notwithstanding the provisions of Directive 2008/95/EC.

2. Without prejudice to Article 6(4), a trade mark the use of which contravenes Article 13(1) which has been applied for, registered, or established by use if that possibility is provided for by the legislation concerned, in good faith within the territory of the Union, before the date on which the application for protection of the designation of origin or geographical indication is submitted to the Commission, may continue to be used and renewed for that product notwithstanding the registration of a designation of origin or geographical indication, provided that no grounds for its invalidity or revocation exist under Council Regulation (EC) No 207/2009 of 26 February 2009 on the Community trade mark [1] or under Directive 2008/95/EC. In such cases, the use of the protected designation of origin or protected geographical indication shall be permitted as well as use of the relevant trade marks.

Article 15

Transitional periods for use of protected designations of origin and protected geographical indications

1. Without prejudice to Article 14, the Commission may adopt implementing acts granting a transitional period of up to five years to enable products originating in a Member State or a third country the designation of which consists of or contains a name that contravenes Article 13(1) to continue to use the designation under which it was marketed on condition that an admissible statement of opposition under Article 49(3) or Article 51 shows that:

(a) the registration of the name would jeopardise the existence of an entirely or partly identical name; or

(b) such products have been legally marketed with that name in the territory concerned for at least five years preceding the date of the publication provided for point (a) of Article 50(2).

Those implementing acts shall be adopted in accordance with the examination procedure referred to in Article 57(2).

2. Without prejudice to Article 14, the Commission may adopt implementing acts extending the transitional period mentioned in paragraph 1 of this Article to 15 years in duly justified cases where it is shown that:

(a) the designation referred to in paragraph 1 of this Article has been in legal use consistently and fairly for at least 25 years before the application for registration was submitted to the Commission;

(b) the purpose of using the designation referred to in paragraph 1 of this Article has not, at any time, been to profit from the reputation of the registered name and it is shown that the consumer has not been nor could have been misled as to the true origin of the product.

Those implementing acts shall be adopted in accordance with the examination procedure referred to in Article 57(2).

3. When using a designation referred to in paragraphs 1 and 2, the indication of country of origin shall clearly and visibly appear on the labelling.

4. To overcome temporary difficulties with the long-term objective of ensuring that all producers in the area concerned comply with the specification, a Member State may grant a transitional period of up to 10 years, with effect from the date on which the application is lodged with the Commission, on condition that the operators concerned have legally marketed the products in question, using the names concerned continuously for at least the five years prior to the lodging of the application to the authorities of the Member State and have made that point in the national opposition procedure referred to in Article 49(3).

The first subparagraph shall apply *mutatis mutandis* to a protected geographical indication or protected designation of origin referring to a geographical area situated in a third country, with the exception of the opposition procedure.

Such transitional periods shall be indicated in the application dossier referred to in Article 8(2).

Article 16

Transitional provisions

1. Names entered in the register provided for in Article 7(6) of Regulation (EC) No 510/2006 shall automatically be entered in the register referred to in Article 11 of this Regulation. The corresponding specifications shall be deemed to be the specifications referred to in Article 7 of this Regulation. Any specific transitional provisions associated with such registrations shall continue to apply.

(1) OJ L 78, 24.3.2009, p. 1.

2. In order to protect the rights and legitimate interests of producers or stakeholders concerned, the Commission shall be empowered to adopt delegated acts, in accordance with Article 56, concerning additional transitional rules.

3. This Regulation shall apply without prejudice to any right of coexistence recognised under Regulation (EC) No 510/2006 in respect of designations of origin and geographical indications, on the one hand, and trade marks, on the other.

TITLE III

TRADITIONAL SPECIALITIES GUARANTEED

Article 17

Objective

A scheme for traditional specialities guaranteed is established to safeguard traditional methods of production and recipes by helping producers of traditional product in marketing and communicating the value-adding attributes of their traditional recipes and products to consumers.

Article 18

Criteria

1. A name shall be eligible for registration as a traditional speciality guaranteed where it describes a specific product or foodstuff that:

(a) results from a mode of production, processing or composition corresponding to traditional practice for that product or foodstuff; or

(b) is produced from raw materials or ingredients that are those traditionally used.

2. For a name to be registered as a traditional speciality guaranteed, it shall:

(a) have been traditionally used to refer to the specific product; or

(b) identify the traditional character or specific character of the product.

3. If it is demonstrated in the opposition procedure under Article 51 that the name is also used in another Member State or in a third country, in order to distinguish comparable products or products that share an identical or similar name, the decision on registration taken in accordance with Article 52(3) may provide that the name of the traditional speciality guaranteed is to be accompanied by the claim 'made following the tradition of' immediately followed by the name of a country or a region thereof.

4. A name may not be registered if it refers only to claims of a general nature used for a set of products, or to claims provided for by particular Union legislation.

5. In order to ensure the smooth functioning of the scheme, the Commission shall be empowered to adopt delegated acts, in accordance with Article 56, concerning further details of the eligibility criteria laid down in this Article.

Article 19

Product specification

1. A traditional speciality guaranteed shall comply with a specification which shall comprise:

(a) the name proposed for registration, in the appropriate language versions;

(b) a description of the product including its main physical, chemical, microbiological or organoleptic characteristics, showing the product's specific character;

(c) a description of the production method that the producers must follow, including, where appropriate, the nature and characteristics of the raw materials or ingredients used, and the method by which the product is prepared; and

(d) the key elements establishing the product's traditional character.

2. In order to ensure that product specifications provide relevant and succinct information, the Commission shall be empowered to adopt delegated acts, in accordance with Article 56, laying down rules which limit the information contained in the specification referred to in paragraph 1 of this Article, where such a limitation is necessary to avoid excessively voluminous applications for registration.

The Commission may adopt implementing acts laying down rules on the form of the specification. Those implementing acts shall be adopted in accordance with the examination procedure referred to in Article 57(2).

Article 20

Content of application for registration

1. An application for registration of a name as a traditional speciality guaranteed referred to in Article 49(2) or (5) shall comprise:

(a) the name and address of the applicant group;

(b) the product specification as provided for in Article 19.

2. An application dossier referred to in Article 49(4) shall comprise:

(a) the elements referred to in paragraph 1 of this Article; and

(b) a declaration by the Member State that it considers that the application lodged by the group and qualifying for the favourable decision meets the conditions of this Regulation and the provisions adopted pursuant thereto.

Article 21

Grounds for opposition

1. A reasoned statement of opposition as referred to in Article 51(2) shall be admissible only if it is received by the Commission before expiry of the time limit and if it:

(a) gives duly substantiated reasons why the proposed registration is incompatible with the terms of this Regulation; or

(b) shows that use of the name is lawful, renowned and economically significant for similar agricultural products or foodstuffs.

2. The criteria referred to in point (b) of paragraph 1 shall be assessed in relation to the territory of the Union.

Article 22

Register of traditional specialities guaranteed

1. The Commission shall adopt implementing acts, without applying the procedure referred to in Article 57(2), establishing and maintaining a publicly accessible updated register of traditional specialties guaranteed recognised under this scheme.

2. The Commission may adopt implementing acts laying down detailed rules on the form and content of the register. Those implementing acts shall be adopted in accordance with the examination procedure referred to in Article 57(2).

Article 23

Names, symbol and indication

1. A name registered as a traditional speciality guaranteed may be used by any operator marketing a product that conforms to the corresponding specification.

2. A Union symbol shall be established in order to publicise the traditional specialities guaranteed.

3. In the case of the products originating in the Union that are marketed under a traditional speciality guaranteed that is registered in accordance with this Regulation, the symbol referred to in paragraph 2 shall, without prejudice to paragraph 4, appear on the labelling. In addition, the name of the product should appear in the same field of vision. The indication 'traditional speciality guaranteed' or the corresponding abbreviation 'TSG' may also appear on the labelling.

The symbol shall be optional on the labelling of traditional specialities guaranteed which are produced outside the Union.

4. In order to ensure that the appropriate information is communicated to the consumer, the Commission shall be empowered to adopt delegated acts, in accordance with Article 56, establishing the Union symbol.

The Commission may adopt implementing acts defining the technical characteristics of the Union symbol and indication, as well as the rules of their use on the products bearing the name of a traditional speciality guaranteed, including as to the appropriate linguistic versions to be used. Those implementing acts shall be adopted in accordance with the examination procedure referred to in Article 57(2).

Article 24

Restriction on use of registered names

1. Registered names shall be protected against any misuse, imitation or evocation, or against any other practice liable to mislead the consumer.

2. Member States shall ensure that sales descriptions used at national level do not give rise to confusion with names that are registered.

3. The Commission may adopt implementing acts laying down rules for the protection of traditional specialities guaranteed. Those implementing acts shall be adopted in accordance with the examination procedure referred to in Article 57(2).

Article 25

Transitional provisions

1. Names registered in accordance with Article 13(2) of Regulation (EC) No 509/2006 shall be automatically entered in the register referred to in Article 22 of this Regulation. The corresponding specifications shall be deemed to be the specifications referred to in Article 19 of this Regulation. Any specific transitional provisions associated with such registrations shall continue to apply.

2. Names registered in accordance with the requirements laid down in Article 13(1) of Regulation (EC) No 509/2006, including those registered pursuant to applications referred to in the second subparagraph of Article 58(1) of this Regulation, may continue to be used under the conditions provided for in Regulation (EC) No 509/2006 until 4 January 2023 unless Member States use the procedure set out in Article 26 of this Regulation.

3. In order to protect the rights and legitimate interests of producers or stakeholders concerned, the Commission shall be empowered to adopt delegated acts, in accordance with Article 56, laying down additional transitional rules.

Article 26

Simplified procedure

1. At the request of a group, a Member State may submit, no later than 4 January 2016, to the Commission names of traditional specialities guaranteed that are registered in accordance with Article 13(1) of Regulation (EC) No 509/2006 and that comply with this Regulation.

Before submitting a name, the Member State shall initiate an opposition procedure as defined in Article 49(3) and (4).

If it is demonstrated in the course of this procedure that the name is also used in reference to comparable products or products that share an identical or similar name, the name may be complemented by a term identifying its traditional or specific character.

A group from a third country may submit such names to the Commission, either directly or through the authorities of the third country.

2. The Commission shall publish the names referred to in paragraph 1 together with the specifications for each such name in the *Official Journal of the European Union* within two months from reception.

3. Articles 51 and 52 shall apply.

4. Once the opposition procedure has finished, the Commission shall, where appropriate, adjust the entries in the register set out in Article 22. The corresponding specifications shall be deemed to be the specifications referred to in Article 19.

TITLE IV

OPTIONAL QUALITY TERMS

Article 27

Objective

A scheme for optional quality terms is established in order to facilitate the communication within the internal market of the value-adding characteristics or attributes of agricultural products by the producers thereof.

Article 28

National Rules

Member States may maintain national rules on optional quality terms which are not covered by this Regulation, provided that such rules comply with Union law.

Article 29

Optional quality terms

1. Optional quality terms shall satisfy the following criteria:

(a) the term relates to a characteristic of one or more categories of products, or to a farming or processing attribute which applies in specific areas;

(b) the use of the term adds value to the product as compared to products of a similar type; and

(c) the term has a European dimension.

2. Optional quality terms that describe technical product qualities with the purpose of putting into effect compulsory marketing standards and are not intended to inform consumers about those product qualities shall be excluded from this scheme.

3. Optional quality terms shall exclude optional reserved terms which support and complement specific marketing standards determined on a sectoral or product category basis.

4. In order to take into account the specific character of certain sectors as well as consumer expectations, the Commission shall be empowered to adopt delegated acts, in accordance with Article 56, laying down detailed rules relating to the criteria referred to in paragraph 1 of this Article.

5. The Commission may adopt implementing acts laying down all measures related to forms, procedures or other technical details, necessary for the application of this Title. Those implementing acts shall be adopted in accordance with the examination procedure referred to in Article 57(2).

6. When adopting delegated and implementing acts in accordance with paragraphs 4 and 5 of this Article, the Commission shall take account of any relevant international standards.

Article 30

Reservation and amendment

1. In order to take account of the expectations of consumers, developments in scientific and technical knowledge, the market situation, and developments in marketing standards and in international standards, the Commission shall be empowered to adopt delegated acts, in accordance with Article 56, reserving an additional optional quality term and laying down its conditions of use.

2. In duly justified cases and in order to take into account the appropriate use of the additional optional quality term, the Commission shall be empowered to adopt delegated acts, in accordance with Article 56, laying down amendments to the conditions of use referred to in paragraph 1 of this Article.

Article 31

Mountain product

1. The term 'mountain product' is established as an optional quality term.

This term shall only be used to describe products intended for human consumption listed in Annex I to the Treaty in respect of which:

(a) both the raw materials and the feedstuffs for farm animals come essentially from mountain areas;

(b) in the case of processed products, the processing also takes place in mountain areas.

2. For the purposes of this Article, mountain areas within the Union are those delimited pursuant to Article 18(1) of Regulation (EC) No 1257/1999. For third-country products, mountain areas include areas officially designated as mountain areas by the third country or that meet criteria equivalent to those set out in Article 18(1) of Regulation (EC) No 1257/1999.

3. In duly justified cases and in order to take into account natural constraints affecting agricultural production in mountain areas, the Commission shall be empowered to adopt delegated acts, in accordance with Article 56, laying down derogations from the conditions of use referred to in paragraph 1 of this Article. In particular, the Commission shall be empowered to adopt a delegated act laying down the conditions under which raw materials or feedstuffs are permitted to come from outside the mountain areas, the conditions under which the processing of products is permitted to take place outside of the mountain areas in a geographical area to be defined, and the definition of that geographical area.

4. In order to take into account natural constraints affecting agricultural production in mountain areas, the Commission shall be empowered to adopt delegated acts, in accordance with Article 56, concerning the establishment of the methods of production, and other criteria relevant for the application of the optional quality term established in paragraph 1 of this Article.

Article 32

Product of island farming

No later than 4 January 2014 the Commission shall present a report to the European Parliament and to the Council on the case for a new term, 'product of island farming'. The term may only be used to describe the products intended for human consumption that are listed in Annex I to the Treaty the raw materials of which come from islands. In addition, for the term to be applied to processed products, processing must also take place on islands in cases where this substantially affects the particular characteristics of the final product.

That report shall, if necessary, be accompanied by appropriate legislative proposals to reserve an optional quality term 'product of island farming'.

Article 33

Restrictions on use

1. An optional quality term may only be used to describe products that comply with the corresponding conditions of use.

2. The Commission may adopt implementing acts laying down rules for the use of optional quality terms. Those implementing acts shall be adopted in accordance with the examination procedure referred to in Article 57(2).

Article 34

Monitoring

Member States shall undertake checks, based on a risk analysis, to ensure compliance with the requirements of this Title and, in the event of breach, shall apply appropriate administrative penalties.

TITLE V

COMMON PROVISIONS

CHAPTER I

Official controls of protected designations of origin, protected geographical indications and traditional specialities guaranteed

Article 35

Scope

The provisions of this Chapter shall apply in respect of the quality schemes set out in Title II and Title III.

Article 36

Designation of competent authority

1. In accordance with Regulation (EC) No 882/2004, Member States shall designate the competent authority or authorities responsible for official controls carried out to verify compliance with the legal requirements related to the quality schemes established by this Regulation.

Procedures and requirements of Regulation (EC) No 882/2004 shall apply *mutatis mutandis* to the official controls carried out to verify compliance with the legal requirement related to the quality schemes for all products covered by Annex I to this Regulation.

2. The competent authorities referred to in paragraph 1 shall offer adequate guarantees of objectivity and impartiality, and shall have at their disposal the qualified staff and resources necessary to carry out their functions.

3. Official controls shall cover:

(a) verification that a product complies with the corresponding product specification; and

(b) monitoring of the use of registered names to describe product placed on the market, in conformity with Article 13 for names registered under Title II and in conformity with Article 24 for names registered under Title III.

Article 37

Verification of compliance with product specification

1. In respect of protected designations of origin, protected geographical indications and traditional specialities guaranteed that designate products originating within the Union, verification of compliance with the product specification, before placing the product on the market, shall be carried out by:

(a) one or more of the competent authorities as referred to in Article 36 of this Regulation; and/or

(b) one or more of the control bodies within the meaning of point (5) of Article 2 of Regulation (EC) No 882/2004 operating as a product certification body.

The costs of such verification of compliance with the specifications may be borne by the operators that are subject to those controls. The Member States may also contribute to these costs.

2. In respect of designations of origin, geographical indications and traditional specialities guaranteed that designate products originating in a third country, the verification of compliance with the specifications before placing the product on the market shall be carried out by:

(a) one or more of the public authorities designated by the third country; and/or

(b) one or more of the product certification bodies.

3. Member States shall make public the name and address of the authorities and bodies referred to paragraph 1 of this Article, and update that information periodically.

The Commission shall make public the name and address of the authorities and bodies referred to in paragraph 2 of this Article and update that information periodically.

4. The Commission may adopt implementing acts, without applying the procedure referred to in Article 57(2), defining the means by which the name and address of product certification bodies referred to in paragraphs 1 and 2 of this Article shall be made public.

Article 38

Surveillance of the use of the name in the market place

Member States shall inform the Commission of the names and addresses of the competent authorities referred to in Article 36. The Commission shall make public the names and addresses of those authorities.

Member States shall carry out checks, based on a risk analysis, to ensure compliance with the requirements of this Regulation and, in the event of breaches, Member States shall take all necessary measures.

Article 39

Delegation by competent authorities to control bodies

1. Competent authorities may delegate, in accordance with Article 5 of Regulation (EC) No 882/2004, specific tasks related to official controls of the quality schemes to one or more control bodies.

2. Such control bodies shall be accredited in accordance with European Standard EN 45011 or ISO/IEC Guide 65 (General requirements for bodies operating product certification systems).

3. Accreditation referred to in paragraph 2 of this Article may only be performed by:

(a) a national accreditation body in the Union in accordance with the provisions of Regulation (EC) No 765/2008; or

(b) an accreditation body outside the Union that is a signatory of a multilateral recognition arrangement under the auspices of the International Accreditation Forum.

Article 40

Planning and reporting of control activities

1. Member States shall ensure that activities for the control of obligations under this Chapter are specifically included in a separate section within the multi-annual national control plans in accordance with Articles 41, 42 and 43 of Regulation (EC) No 882/2004.

2. The annual reports concerning the control of the obligations established by this Regulation shall include a separate section comprising the information laid down in Article 44 of Regulation (EC) No 882/2004.

CHAPTER II

Exceptions for certain prior uses

Article 41

Generic terms

1. Without prejudice to Article 13, this Regulation shall not affect the use of terms that are generic in the Union, even if the generic term is part of a name that is protected under a quality scheme.

2. To establish whether or not a term has become generic, account shall be taken of all relevant factors, in particular:

(a) the existing situation in areas of consumption;

(b) the relevant national or Union legal acts.

3. In order to fully protect the rights of interested parties, the Commission shall be empowered to adopt delegated acts, in accordance with Article 56, laying down additional rules for determining the generic status of terms referred to in paragraph 1 of this Article.

Article 42

Plant varieties and animal breeds

1. This Regulation shall not prevent the placing on the market of products the labelling of which includes a name or term protected or reserved under a quality scheme described in Title II, Title III, or Title IV that contains or comprises the name of a plant variety or animal breed, provided that the following conditions are met:

(a) the product in question comprises or is derived from the variety or breed indicated;

(b) consumers are not misled;

(c) the usage of the name of the variety or breed name constitutes fair competition;

(d) the usage does not exploit the reputation of the protected term; and

(e) in the case of the quality scheme described in Title II, production and marketing of the product had spread beyond its area of origin prior to the date of application for registration of the geographical indication.

2. In order to further clarify the extent of rights and freedoms of food business operators to use the name of a plant variety or of an animal breed referred to in paragraph 1 of this Article, the Commission shall be empowered to adopt delegated acts, in accordance with Article 56, concerning rules for determining the use of such names.

Article 43

Relation to intellectual property

The quality schemes described in Titles III and IV shall apply without prejudice to Union rules or to those of Member States governing intellectual property, and in particular to those concerning designations of origin and geographical indications and trade marks, and rights granted under those rules.

CHAPTER III

Quality scheme indications and symbols and role of producers

Article 44

Protection of indications and symbols

1. Indications, abbreviations and symbols referring to the quality schemes may only be used in connection with products produced in conformity with the rules of the quality scheme to which they apply. This applies in particular to the following indications, abbreviations and symbols:

(a) 'protected designation of origin', 'protected geographical indication', 'geographical indication', 'PDO', 'PGI', and the associated symbols, as provided for in Title II;

(b) 'traditional speciality guaranteed', 'TSG', and the associated symbol, as provided for in Title III;

(c) 'mountain product', as provided for in Title IV.

2. In accordance with Article 5 of Regulation (EC) No 1290/2005, the European Agricultural Fund for Rural Development (EAFRD) may, on the initiative of the Commission or on its behalf, finance, on a centralised basis, administrative support concerning the development, preparatory work, monitoring, administrative and legal support, legal defence, registration fees, renewal fees, trade mark watching fees, litigation fees and any other related measure required to protect the use of the indications, abbreviations and symbols referring to the quality schemes from misuse, imitation, evocation or any other practice liable to mislead the consumer, within the Union and in third countries.

3. The Commission shall adopt implementing acts laying down rules for the uniform protection of the indications, abbreviations and symbols referred to in paragraph 1 of this Article. Those implementing acts shall be adopted in accordance with the examination procedure referred to in Article 57(2).

Article 45

Role of groups

1. Without prejudice to specific provisions on producer organisations and inter-branch organisations as laid down in Regulation (EC) No 1234/2007, a group is entitled to:

(a) contribute to ensuring that the quality, reputation and authenticity of their products are guaranteed on the market by monitoring the use of the name in trade and, if necessary, by informing competent authorities as referred to in Article 36, or any other competent authority within the framework of Article 13(3);

(b) take action to ensure adequate legal protection of the protected designation of origin or protected geographical indication and of the intellectual property rights that are directly connected with them;

(c) develop information and promotion activities aiming at communicating the value-adding attributes of the product to consumers;

(d) develop activities related to ensuring compliance of a product with its specification;

(e) take action to improve the performance of the scheme, including developing economic expertise, carrying out economic analyses, disseminating economic information on the scheme and providing advice to producers;

(f) take measures to enhance the value of products and, where necessary, take steps to prevent or counter any measures which are, or risk being, detrimental to the image of those products.

2. Member States may encourage the formation and functioning of groups on their territories by administrative means. Moreover, Member States shall communicate to the Commission the name and address of the groups referred to in point 2 of Article 3. The Commission shall make this information public.

Article 46

Right to use the schemes

1. Member States shall ensure that any operator complying with the rules of a quality scheme set out in Titles II and III is entitled to be covered by the verification of compliance established pursuant to Article 37.

2. Operators who prepare and store a product marketed under the traditional speciality guaranteed, protected designation of origin or protected geographical indication schemes or who place such products on the market shall also be subject to the controls laid down in Chapter I of this Title.

3. Member States shall ensure that operators willing to adhere to the rules of a quality scheme set out in Titles III and IV are able to do so and do not face obstacles to participation that are discriminatory or otherwise not objectively founded.

Article 47

Fees

Without prejudice to Regulation (EC) No 882/2004 and in particular the provisions of Chapter VI of Title II thereof, Member States may charge a fee to cover their costs of managing the quality schemes, including those incurred in processing applications, statements of opposition, applications for amendments and requests for cancellations provided for in this Regulation.

CHAPTER IV

Application and registration processes for designations of origin, geographical indications, and traditional specialities guaranteed

Article 48

Scope of application processes

The provisions of this Chapter shall apply in respect of the quality schemes set out in Title II and Title III.

Article 49

Application for registration of names

1. Applications for registration of names under the quality schemes referred to in Article 48 may only be submitted by groups who work with the products with the name to be registered. In the case of a 'protected designations of origin' or 'protected geographical indications' name that designates a trans-border geographical area or in the case of a 'traditional specialities guaranteed' name, several groups from different Member States or third countries may lodge a joint application for registration.

A single natural or legal person may be treated as a group where it is shown that both of the following conditions are fulfilled:

(a) the person concerned is the only producer willing to submit an application;

(b) with regard to protected designations of origin and protected geographical indications, the defined geographical area possesses characteristics which differ appreciably from those of neighbouring areas or the characteristics of the product are different from those produced in neighbouring areas.

2. Where the application under the scheme set out in Title II relates to a geographical area in a Member State, or where an application under the scheme set out in Title III is prepared by a group established in a Member State, the application shall be addressed to the authorities of that Member State.

The Member State shall scrutinise the application by appropriate means in order to check that it is justified and meets the conditions of the respective scheme.

3. As part of the scrutiny referred to in the second subparagraph of paragraph 2 of this Article, the Member State shall initiate a national opposition procedure that ensures adequate publication of the application and that provides for a reasonable period within which any natural or legal person having a legitimate interest and established or resident on its territory may lodge an opposition to the application.

The Member State shall examine the admissibility of oppositions received under the scheme set out in Title II in the light of the criteria referred to in Article 10(1), or the admissibility of oppositions received under the scheme set out in Title III in the light of the criteria referred to in Article 21(1).

4. If, after assessment of any opposition received, the Member State considers that the requirements of this Regulation are met, it may take a favourable decision and lodge an application dossier with the Commission. It shall in such case inform the Commission of admissible oppositions received from a natural or legal person that have legally marketed the products in question, using the names concerned continuously for at least five years preceding the date of the publication referred to in paragraph 3.

The Member State shall ensure that its favourable decision is made public and that any natural or legal person having a legitimate interest has an opportunity to appeal.

The Member State shall ensure that the version of the product specification on which its favourable decision is based, is published, and shall provide electronic access to the product specification.

With reference to protected designations of origin and protected geographical indications, the Member State shall also ensure adequate publication of the version of the product specification on which the Commission takes its decision pursuant to Article 50(2).

5. Where the application under the scheme set out in Title II relates to a geographical area in a third country, or where an application under the scheme set out in Title III is prepared by a group established in a third country, the application shall be lodged with the Commission, either directly or via the authorities of the third country concerned.

6. The documents referred to in this Article which are sent to the Commission shall be in one of the official languages of the Union.

7. In order to facilitate the application process, the Commission shall be empowered to adopt delegated acts, in accordance with Article 56, defining the rules for carrying out the national objection procedure for joint applications concerning more than one national territory and complementing the rules of the application process.

The Commission may adopt implementing acts laying down detailed rules on procedures, form and presentation of applications, including for applications concerning more than one national territory. Those implementing acts shall be adopted in accordance with the examination procedure referred to in Article 57(2).

Article 50

Scrutiny by the Commission and publication for opposition

1. The Commission shall scrutinise by appropriate means any application that it receives pursuant to Article 49, in order to check that it is justified and that it meets the conditions of the respective scheme. This scrutiny should not exceed a period of six months. Where this period is exceeded, the Commission shall indicate in writing to the applicant the reasons for the delay.

The Commission shall, at least each month, make public the list of names for which registration applications have been submitted to it, as well as their date of submission.

2. Where, based on the scrutiny carried out pursuant to the first subparagraph of paragraph 1, the Commission considers that the conditions laid down in this Regulation are fulfilled, it shall publish in the *Official Journal of the European Union*:

(a) for applications under the scheme set out in Title II, the single document and the reference to the publication of the product specification;

(b) for applications under the scheme set out in Title III, the specification.

Article 51

Opposition procedure

1. Within three months from the date of publication in the *Official Journal of the European Union*, the authorities of a Member State or of a third country, or a natural or legal person having a legitimate interest and established in a third country may lodge a notice of opposition with the Commission.

Any natural or legal person having a legitimate interest, established or resident in a Member State other than that from which the application was submitted, may lodge a notice of opposition with the Member State in which it is established within a time limit permitting an opposition to be lodged pursuant to the first subparagraph.

A notice of opposition shall contain a declaration that the application might infringe the conditions laid down in this Regulation. A notice of opposition that does not contain this declaration is void.

The Commission shall forward the notice of opposition to the authority or body that lodged the application without delay.

2. If a notice of opposition is lodged with the Commission and is followed within two months by a reasoned statement of opposition, the Commission shall check the admissibility of this reasoned statement of opposition.

3. Within two months after the receipt of an admissible reasoned statement of opposition, the Commission shall invite the authority or person that lodged the opposition and the authority or body that lodged the application to engage in appropriate consultations for a reasonable period that shall not exceed three months.

The authority or person that lodged the opposition and the authority or body that lodged the application shall start such appropriate consultations without undue delay. They shall provide each other with the relevant information to assess whether the application for registration complies with the conditions of this Regulation. If no agreement is reached, this information shall also be provided to the Commission.

At any time during these three months, the Commission may, at the request of the applicant extend the deadline for the consultations by a maximum of three months.

4. Where, following the appropriate consultations referred to in paragraph 3 of this Article, the details published in accordance with Article 50(2) have been substantially amended, the Commission shall repeat the scrutiny referred to in Article 50.

5. The notice of opposition, the reasoned statement of opposition and the related documents which are sent to the Commission in accordance with paragraphs 1 to 4 of this Article shall be in one of the official languages of the Union.

6. In order to establish clear procedures and deadlines for opposition, the Commission shall be empowered to adopt delegated acts, in accordance with Article 56, complementing the rules of the opposition procedure.

The Commission may adopt implementing acts laying down detailed rules on procedures, form and presentation of the oppositions. Those implementing acts shall be adopted in accordance with the examination procedure referred to in Article 57(2).

Article 52

Decision on registration

1. Where, on the basis of the information available to the Commission from the scrutiny carried out pursuant to the first subparagraph of Article 50(1), the Commission considers that the conditions for registration are not fulfilled, it shall adopt implementing acts rejecting the application. Those implementing acts shall be adopted in accordance with the examination procedure referred to in Article 57(2).

2. If the Commission receives no notice of opposition or no admissible reasoned statement of opposition under Article 51, it shall adopt implementing acts, without applying the procedure referred to in Article 57(2), registering the name.

3. If the Commission receives an admissible reasoned statement of opposition, it shall, following the appropriate consultations referred to in Article 51(3), and taking into account the results thereof, either:

(a) if an agreement has been reached, register the name by means of implementing acts adopted without applying the procedure referred to in Article 57(2), and, if necessary, amend the information published pursuant to Article 50(2) provided such amendments are not substantial; or

(b) if an agreement has not been reached, adopt implementing acts deciding on the registration. Those implementing acts shall be adopted in accordance with the examination procedure referred to in Article 57(2).

4. Acts of registration and decisions on rejection shall be published in the *Official Journal of the European Union*.

Article 53

Amendment to a product specification

1. A group having a legitimate interest may apply for approval of an amendment to a product specification.

Applications shall describe and give reasons for the amendments requested.

2. Where the amendment involves one or more amendments to the specification that are not minor, the amendment application shall follow the procedure laid down in Articles 49 to 52.

However, if the proposed amendments are minor, the Commission shall approve or reject the application. In the event of the approval of amendments implying a modification of the elements referred to in Article 50(2), the Commission shall publish those elements in the *Official Journal of the European Union*.

For an amendment to be regarded as minor in the case of the quality scheme described in Title II, it shall not:

(a) relate to the essential characteristics of the product;

(b) alter the link referred to in point (f)(i) or (ii) of Article 7(1);

(c) include a change to the name, or to any part of the name of the product;

(d) affect the defined geographical area; or

(e) represent an increase in restrictions on trade in the product or its raw materials.

For an amendment to be regarded as minor in the case of the quality scheme described in Title III, it shall not:

(a) relate to the essential characteristics of the product;

(b) introduce essential changes to the production method; or

(c) include a change to the name, or to any part of the name of the product.

The scrutiny of the application shall focus on the proposed amendment.

3. In order to facilitate the administrative process of an amendment application, including where the amendment does not involve any change to the single document and where it concerns a temporary change in the specification resulting from the imposition of obligatory sanitary or phytosanitary measures by the public authorities, the Commission shall be empowered to adopt delegated acts, in accordance with Article 56, complementing the rules of the amendment application process.

The Commission may adopt implementing acts laying down detailed rules on procedures, form and presentation of an amendment application. Those implementing acts shall be adopted in accordance with the examination procedure referred to in Article 57(2).

Article 54

Cancellation

1. The Commission may, on its own initiative or at the request of any natural or legal person having a legitimate interest, adopt implementing acts to cancel the registration of a protected designation of origin or of a protected geographical indication or of a traditional speciality guaranteed in the following cases:

(a) where compliance with the conditions of the specification is not ensured;

(b) where no product is placed on the market under the traditional speciality guaranteed, the protected designation of origin or the protected geographical indication for at least seven years.

The Commission may, at the request of the producers of product marketed under the registered name, cancel the corresponding registration.

Those implementing acts shall be adopted in accordance with the examination procedure referred to in Article 57(2).

2. In order to ensure legal certainty that all parties have the opportunity to defend their rights and legitimate interests, the Commission shall be empowered to adopt delegated acts, in accordance with Article 56 complementing the rules regarding the cancellation process.

The Commission may adopt implementing acts laying down detailed rules on procedures and form of the cancellation process, as well as on the presentation of the requests referred to in paragraph 1 of this Article. Those implementing acts shall be adopted in accordance with the examination procedure referred to in Article 57(2).

TITLE VI

PROCEDURAL AND FINAL PROVISIONS

CHAPTER I

Local farming and direct sales

Article 55

Reporting on local farming and direct sales

No later than 4 January 2014 the Commission shall present a report to the European Parliament and to the Council on the case for a new local farming and direct sales labelling scheme to assist producers in marketing their produce locally. That report shall focus on the ability of the farmer to add value to his produce through the new label, and should take into account other criteria, such as the possibilities of reducing carbon emissions and waste through short production and distribution chains.

That report shall, if necessary, be accompanied by appropriate legislative proposals on the creation of a local farming and direct sales labelling scheme.

CHAPTER II

Procedural rules

Article 56

Exercise of the delegation

1. The power to adopt the delegated acts is conferred on the Commission subject to the conditions laid down in this Article.

2. The power to adopt delegated acts referred to in the second subparagraph of Article 2(1), Article 5(4), the first subparagraph of Article 7(2), the first subparagraph of Article 12(5), Article 16(2), Article 18(5), the first subparagraph of Article 19(2), the first subparagraph of Article 23(4), Article 25(3), Article 29(4), Article 30, Article 31(3) and (4), Article 41(3), Article 42(2), the first subparagraph of Article 49(7), the first subparagraph of Article 51(6), the first subparagraph of Article 53(3) and the first subparagraph of Article 54(2) shall be conferred on the Commission for a period of five years from 3 January 2013. The Commission shall draw up a report in respect of the delegation of power not later than nine months before the end of the five-year period. The delegation of power shall be tacitly extended for periods of an identical duration, unless the European Parliament or the Council opposes such extension not later than three months before the end of each period.

3. The delegation of power referred to in the second subparagraph of Article 2(1), Article 5(4), the first subparagraph of Article 7(2), the first subparagraph of Article 12(5), Article 16(2), Article 18(5), the first subparagraph of Article 19(2), the first subparagraph of Article 23(4), Article 25(3), Article 29(4), Article 30, Article 31(3) and (4), Article 41(3), Article 42(2), the first subparagraph of Article 49(7), the first subparagraph of Article 51(6), the first subparagraph of Article 53(3) and the first subparagraph of Article 54(2) may be revoked at any time by the European Parliament or by the Council. A decision to revoke shall put an end to the delegation of the power specified in that decision. It shall take effect the day following the publication of the decision in the *Official Journal of the European Union* or at a later date specified therein. It shall not affect the validity of any delegated acts already in force.

4. As soon as it adopts a delegated act, the Commission shall notify it simultaneously to the European Parliament and to the Council.

5. A delegated act adopted pursuant to the second subparagraph of Article 2(1), Article 5(4), the first subparagraph of Article 7(2), the first subparagraph of Article 12(5), Article 16(2), Article 18(5), the first subparagraph of Article 19(2), the first subparagraph of Article 23(4), Article 25(3), Article 29(4), Article 30, Article 31(3) and (4), Article 41(3), Article 42(2), the first subparagraph of Article 49(7), the first subparagraph of Article 51(6), the first subparagraph of Article 53(3) and the first subparagraph of Article 54(2) shall enter into force only if no objection has been expressed either by the European Parliament or the Council within a period of two months of notification of that act to the European Parliament and the Council or if, before the expiry of that period, the European Parliament and the Council have both informed the Commission that they will not object. That period shall be extended by two months at the initiative of the European Parliament or of the Council.

Article 57

Committee procedure

1. The Commission shall be assisted by the Agricultural Product Quality Policy Committee. That committee shall be a committee within the meaning of Regulation (EU) No 182/2011.

2. Where reference is made to this paragraph, Article 5 of Regulation (EU) No 182/2011 shall apply.

Where the committee delivers no opinion, the Commission shall not adopt the draft implementing act and the third subparagraph of Article 5(4) of Regulation (EU) No 182/2011 shall apply.

CHAPTER III

Repeal and final provisions

Article 58

Repeal

1. Regulations (EC) No 509/2006 and (EC) No 510/2006 are hereby repealed.

However, Article 13 of Regulation (EC) No 509/2006 shall continue to apply in respect of applications concerning products falling outside the scope of Title III of this Regulation, received by the Commission prior to the date of entry into force of this Regulation.

2. References to the repealed Regulations shall be construed as references to this Regulation and be read in accordance with the correlation table in Annex II to this Regulation.

Article 59

Entry into force

This Regulation shall enter into force on the twentieth day following that of its publication in the *Official Journal of the European Union*.

However, Article 12(3) and Article 23(3) shall apply from 4 January 2016, without prejudice to products already placed on the market before that date.

This Regulation shall be binding in its entirety and directly applicable in all Member States.

Done at Strasbourg, 21 November 2012.

For the European Parliament	*For the Council*
The President	*The President*
M. SCHULZ	A. D. MAVROYIANNIS

L 343/24 EN Official Journal of the European Union 14.12.2012

ANNEX I

AGRICULTURAL PRODUCTS AND FOODSTUFFS REFERRED TO IN ARTICLE 2(1)

I. Designations of Origin and Geographical indications

— beer,

— chocolate and derived products,

— bread, pastry, cakes, confectionery, biscuits and other baker's wares,

— beverages made from plant extracts,

— pasta,

— salt,

— natural gums and resins,

— mustard paste,

— hay,

— essential oils,

— cork,

— cochineal,

— flowers and ornamental plants,

— cotton,

— wool,

— wicker,

— scutched flax,

— leather,

— fur,

— feather.

II. Traditional specialities guaranteed

— prepared meals,

— beer,

— chocolate and derived products,

— bread, pastry, cakes, confectionery, biscuits and other baker's wares,

— beverages made from plant extracts,

— pasta,

— salt.

ANNEX II

CORRELATION TABLE REFERRED TO IN ARTICLE 58(2)

Regulation (EC) No 509/2006	This Regulation
Article 1(1)	Article 2(1)
Article 1(2)	Article 2(3)
Article 1(3)	Article 2(4)
Article 2(1), point (a)	Article 3, point (5)
Article 2(1), point (b)	Article 3, point (3)
Article 2(1), point (c)	—
Article 2(1), point (d)	Article 3, point (2)
Article 2(2), first to third subparagraph	—
Article 2(2), fourth subparagraph	—
Article 3	Article 22(1)
Article 4(1), first subparagraph	Article 18(1)
Article 4(2)	Article 18(2)
Article 4(3), first subparagraph	—
Article 4(3), second subparagraph	Article 18(4)
Article 5(1)	Article 43
Article 5(2)	Article 42(1)
Article 6(1)	Article 19(1)
Article 6(1), point (a)	Article 19(1), point (a)
Article 6(1), point (b)	Article 19(1), point (b)
Article 6(1), point (c)	Article 19(1), point (c)
Article 6(1), point (d)	—
Article 6(1), point (e)	Article 19(1), point (d)
Article 6(1), point (f)	—
Article 7(1) and (2)	Article 49(1)
Article 7(3), points (a) and (b)	Article 20(1), points (a) and (b)
Article 7(3), point (c)	—
Article 7(3), point (d)	—
Article 7(4)	Article 49(2)

Regulation (EC) No 509/2006	This Regulation
Article 7(5)	Article 49(3)
Article 7(6), points (a), (b) and (c)	Article 49(4)
Article 7(6), point (d)	Article 20(2)
Article 7(7)	Article 49(5)
Article 7(8)	Article 49(6)
Article 8(1)	Article 50(1)
Article 8(2), first subparagraph	Article 50(2), point (b)
Article 8(2), second subparagraph	Article 52(1)
Article 9(1) and (2)	Article 51(1)
Article 9(3)	Article 21(1) and (2)
Article 9(4)	Article 52(2)
Article 9(5)	Article 52(3) and (4)
Article 9(6)	Article 51(5)
Article 10	Article 54
Article 11	Article 53
Article 12	Article 23
Article 13(1)	—
Article 13(2)	—
Article 13(3)	—
Article 14(1)	Article 36(1)
Article 14(2)	Article 46(1)
Article 14(3)	Article 37(3), second subparagraph
Article 15(1)	Article 37(1)
Article 15(2)	Article 37(2)
Article 15(3)	Article 39(2)
Article 15(4)	Article 36(2)
Article 16	—
Article 17(1) and (2)	Article 24(1)
Article 17(3)	Article 24(2)

Regulation (EC) No 509/2006	This Regulation
Article 18	Article 57
Article 19(1), point (a)	—
Article 19(1), point (b)	Article 49(7), second subparagraph
Article 19(1), point (c)	Article 49(7), first subparagraph
Article 19(1), point (d)	Article 22(2)
Article 19(1), point (e)	Article 19(1), point (f)
Article 51(6)	Article 54(1)
Article 19(1), point (g)	Article 23(4)
Article 19(1), point (h)	—
Article 19(1), point (i)	—
Article 19(2)	Article 25(1)
Article 19(3), point (a)	—
Article 19(3), point (b)	Article 25(2)
Article 20	Article 47
Article 21	Article 58
Article 22	Article 59
Annex I	Annex I (Part II)

Regulation (EC) No 510/2006	This Regulation
Article 1(1)	Article 2(1) and (2)
Article 1(2)	Article 2(3)
Article 1(3)	Article 2(4)
Article 2	Article 5
Article 3(1), first subparagraph	Article 6(1)
Article 3(1), second and third subparagraph	Article 41(1), (2) and (3)
Article 3(2), (3) and (4)	Article 6(2), (3) and (4)
Article 4	Article 7
Article 5(1)	Article 3, point (2), and Article 49(1)

Regulation (EC) No 510/2006	This Regulation
Article 5(2)	Article 49(1)
Article 5(3)	Article 8(1)
Article 5(4)	Article 49(2)
Article 5(5)	Article 49(3)
Article 5(6)	Article 9
Article 5(7)	Article 8(2)
Article 5(8)	—
Article 5(9), first subparagraph	—
Article 5(9), second subparagraph	Article 49(5)
Article 5(10)	Article 49(6)
Article 5(11)	—
Article 6(1), first subparagraph	Article 50(1)
Article 6(2), first subparagraph	Article 50(2), point (a)
Article 6(2), second subparagraph	Article 52(1)
Article 7(1)	Article 51(1), first subparagraph
Article 7(2)	Article 51(1), second subparagraph
Article 7(3)	Article 10
Article 7(4)	Article 52(2) and (4)
Article 7(5)	Article 51(3) and Article 52(3) and (4)
Article 7(6)	Article 11
Article 7(7)	Article 51(5)
Article 8	Article 12
Article 9	Article 53
Article 10(1)	Article 36(1)
Article 10(2)	Article 46(1)
Article 10(3)	Article 37(3), second subparagraph
Article 11(1)	Article 37(1)
Article 11(2)	Article 37(2)

14.12.2012 EN Official Journal of the European Union L 343/29

Regulation (EC) No 510/2006	This Regulation
Article 11(3)	Article 39(2)
Article 11(4)	Article 36(2)
Article 12	Article 54
Article 13(1)	Article 13(1)
Article 13(2)	Article 13(2)
Article 13(3)	Article 15(1)
Article 13(4)	Article 15(2)
Article 14	Article 14
Article 15	Article 57
Article 16, point (a)	Article 5(4), second subparagraph
Article 16, point (b)	—
Article 16, point (c)	—
Article 16, point (d)	Article 49(7)
Article 16, point (e)	—
Article 16, point (f)	Article 51(6)
Article 16, point (g)	Article 12(7)
Article 16, point (h)	—
Article 16, point (i)	Article 11(3)
Article 16, point (j)	—
Article 16, point (k)	Article 54(2)
Article 17	Article 16
Article 18	Article 47
Article 19	Article 58
Article 20	Article 59
Annex I and Annex II	Annex I (Part I)

Index